GLOBAL CITIZENSHIP
FROM SOCIAL ANALYSIS TO SOCIAL ACTION

Centennial College

TOP HAT   **NELSON**

TOP HAT

COPYRIGHT © 2015
by Tophatmonocle Corp.

Printed and bound in Canada
21 22 23   23 22 21 20

For more information contact
Tophatmonocle Corp.,
151 Bloor Street West, Suite 200,
Toronto, Ontario, M5S 1S4.
Or you can visit tophat.com

ISBN-13: 978-0-17-658897-7
ISBN-10: 0-17-658897-3

Cover Image:
Shutterstock 145733135
(hand tree) © Cienpies
Design/Shutterstock

Centennial College and its Board of Governors value and embrace diversity, equity and inclusion as fundamental to our mission to educate students for career success within a context of global citizenship and social justice.

We recognize that historical and persistent inequities and barriers to equitable participation exist and are well documented in society and within the college.

We believe individual and systemic biases contribute to the marginalization of designated groups. These biases include race, sex, gender, sexual orientation, age, disability, ancestry, nationality, place of origin, colour, ethnicity, culture, linguistic origin, citizenship, creed (religion, faith), marital status, socio-economic class, family status, receipt of public assistance or record of offence. We acknowledge that resolving First Nations sovereignty issues is fundamental to pursuing equity and social justice within Canada.

We acknowledge the richness and diversity of the community we serve. As our community has evolved, and our staff and student population have changed, we have implemented policies and practices to address issues of inclusion. In moving forward, we will build on this work to embed commitment to diversity, equity and inclusion in every aspect of what we do.

Our Guiding Principles

We believe social justice requires that we value diversity, equity and inclusion. We believe that the principles and practices of diversity, equity and inclusion strengthen the social and economic development, growth and well-being of our student population, our employees, and our local and international communities.

We uphold our social responsibility to contribute to a society that is equitable, fair and just. In accordance with our mission, vision and values, we will demonstrate leadership in eliminating barriers, and implementing and promoting diversity through our Academic Framework, policies, special initiatives and proactive measures.

We are committed to eliminating all forms of harassment and discrimination. We will prevent, remedy and redress these inequities. We will create an environment of inclusion in teaching, learning, employment and support services so we can fully serve our communities and prepare students to excel in the workplace and in society.

We will be accountable for the changes we need to make. We will continue to comply with existing federal and provincial legislative requirements. We will continue to develop and implement goals, policies, competencies and special initiatives founded upon principles of social justice to promote equity and inclusion. We will collect data to track progress and regularly evaluate the effectiveness of the initiatives we undertake, and we will communicate the outcomes to our community.

Our Commitment

A safe, secure, inclusive and accessible environment for learning, teaching and working

Centennial College will be free from discrimination, harassment and hate. We will fully support the right of everyone to study, participate and work with dignity in an environment of mutual respect. We will include and respect the abilities, experiences, perspectives and contributions of our students, our employees, our partners and our communities.

Curriculum and instruction that reflect diversity and promote equity and inclusion

Our curriculum and instruction will draw on a variety of knowledge, perspectives and experiences. Our teaching and learning will help students recognize different forms of discrimination and understand the factors that cause inequity in society. Through our commitment to global citizenship and social justice, we will prepare students with the skills and knowledge to challenge unjust practices and build positive human relationships in an increasingly diverse society.

Equitable and accessible opportunities for student success

We will identify and remove institutional barriers that prevent access and impede student success. Our teaching and support services will demonstrate equity and inclusion. We will provide transformative and inclusive curriculum that will help students attain academic excellence and positive social and career outcomes.

Building knowledge and evaluating effectiveness

We will ensure we are knowledgeable about diversity, equity and inclusion. We will critically analyze and research current practices. We will evaluate our effectiveness by tracking our progress, analyzing what is working well and determining how we can best improve.

Human Resource Management systems, policies and practices that reflect diversity and promote equity and inclusion

We will implement bias- and barrier-free recruitment, selection, hiring and promotion at all levels. We will ensure that our employees' skills and knowledge are respected, valued and used appropriately. We will provide equitable opportunities for professional development and advancement for all employees.

Training and staff development in equity and diversity

We will provide ongoing training and staff development to build understanding and ensure that equity and inclusion are central to the work we do. We will recognize and reward initiatives that support diversity, equity and inclusion.

Accessible and inclusive college communication

We will reflect diversity in communications that promote Centennial College, our programs, services and curriculum. We will ensure that college communication is respectful, and that our information is accessible and widely available.

Strategic engagement with diverse communities

We will undertake strategic outreach to develop meaningful relationships with diverse communities. We will engage with these communities and encourage their fullest contribution to, and participation in, our activities and consultations.

Relationships and partnerships that align with our mission, vision and values

We will actively seek relationships that enhance our values and offer domestic and international opportunities to prepare our students to work effectively and successfully in a global and diverse marketplace. We will ensure that our contractual relationships with businesses and organizations comply with our standards of equity, human rights and fairness.

Committing financial and human resources to promote diversity, equity and inclusion

We will provide resources to support the work of our employees, our students and our partners in promoting diversity, equity and inclusion.

In recent years, there has been a growing concern in colleges and universities about the issue of citizenship. As Martha Nussbaum (1997) suggests, our colleges are not just producing students; they are also producing citizens, and as such, "we must ask what a good citizen of the present day should be and should know" (p. 8). This was the very question that Centennial College was encouraged to explore when President Ann Buller joined the College in 2004. The dialogue began amongst the College community and resulted in the development of a Signature Learning Experience (SLE) that would focus on the concepts of global citizenship and equity and that would be incorporated into all areas of academic and College life.

At Centennial, our administrators, faculty, and staff are committed to educating our students to achieve a broader understanding of the social issues that challenge us in the 21st century including "human rights/equity, peace/justice, environment/ energy/technology and poverty" (Singh, 2009, p. xi). Learning about such issues as injustice, power, and diversity—about the globalized world in which we live—is a matter of necessity because students will inevitably be confronted with these issues and will be required to think critically to seek resolution. As Nussbaum (1997) indicates, "the new emphasis on 'diversity' in college...curricula is above all a way of grappling with the altered requirements of citizenship, an attempt to produce adults who can function as citizens not just of some local region or group but also, and more importantly, as citizens of a complex interlocking world" (p. 6).

Centennial College's SLE provides students with the skills, knowledge, and education that will allow them to achieve a greater global consciousness and to strive towards change. These skills are instrumental not only in their personal life, but also in their professional life as public and private sector employers have indicated that productivity and revenues are compromised when conflicts related to discrimination and harassment occur (Centennial College, 2010). Employers seek graduates with subject matter expertise and an understanding of global issues that contribute to their leadership and interpersonal skills. By creating conditions of inclusion in our classroom, we not only embark on practising the principles of diversity and equity, we also "strengthen the social and economic development, growth and well-being of our student population, our employees, and local and international communities" (Centennial College, 2014). Hence, Centennial College continues to work on a number of programs and initiatives that include:

- a core general education course, Global Citizenship: From Social Analysis to Social Action (GNED 500);
- the Global Citizenship and Equity Learning Experiences (GCELEs): Global and domestic service learning opportunities for students to develop leadership skills and work to create positive social change;
- the inclusion of global citizenship and equity learning outcomes in all programs;
- faculty and staff development learning opportunities;
- current hiring practices that embrace equity and diversity;
- portfolio learning to document individual growth and development; and
- the Institute for Global Citizenship and Equity, which focuses on research and action in the area of global citizenship and equity in addition to publishing both the *Journal of Global Citizenship & Equity Education* and the *Global Citizenship Digest* magazine.

One of the key initiatives of the SLE is our General Education course titled *Global Citizenship: From Social Analysis to Social Action* (GNED 500), which was added to all Centennial's postsecondary programs in 2007 and is a requirement for graduation. GNED 500 provides students with critical pedagogy to engage in discussions on issues relating to social, political, economic, and cultural injustice so that they can reflect on ways in which positive change can be influenced. Students also develop communication, analytical, and conflict resolution skills and obtain a broader knowledge of global issues.

During the search for an appropriate textbook that would cover the course learning outcomes, it became evident that a comprehensive text was not available. Therefore, the General Education and Liberal Studies department faculty wrote the first iteration of the GNED 500 textbook in 2010. Now in its second iteration, this textbook expands on the key theoretical concepts, historical events, and current local and global issues that foster participatory and inclusive learning in the class-room.

The champions of this textbook are the faculty who have worked tirelessly on its production. Their devotion and contribution were evident by the countless number of hours spent on writing and rewriting drafts of the chapters, on engaging in lively discussions and often heated debates, on raising difficult questions and searching collaboratively for solutions, and on motivating one another to do the best work possible. Through this process of writing the textbook, faculty re-enacted the very core values of the course: collaboration, respect, and exchange. As the Chair of the department, I would like to express my sincere gratitude to Rosina Agye-pong, Philip Alailabo, Paula Anderton, Selom Chapman-Nyaho, Agnes Gajewski, Moreen Jones Weekes, Athanasios Tom Kokkinias, Sabrina Malik, Cara Naiman, Jared Purdy, Chet Singh, and Alia Somani for their work on this textbook. Special thanks to Alia Somani for working with the writers on content editing and Cara Naiman for copy editing of the entire textbook. Also, special thanks to the following students, Karen Steele, Shana Isaac Enosse, and Rudi-Marie Blackman for providing extensive feedback on some of the chapters in this book. Finally, I would like to express my appreciation to Nelson Education for working closely with the team to ensure that all aspects of the textbook met publication timelines.

Meera Mather
Chair, General Education and Liberal Studies
School of Advancement
Centennial College

REFERENCES

Centennial College. (2008). *Global citizenship: From social analysis to social action* (2nd ed.). Toronto: Pearson Custom Publishing.

Centennial College. (2010). Leadership and inclusion program advisory committee minutes. (Unpublished document).

Centennial College. (2014). Statement of diversity. Retrieved from https://www1 .centennialcollege.ca/AboutCentennial/Diversity

Nussbaum, M. C. (1997). *Cultivating humanity: A classical defense of reform in liberal education*. Cambridge: Harvard University Press.

Singh. C. (2008). Introduction: Becoming socially literate. In Centennial College, *Global citizenship: From social analysis to social action* (2nd ed., p. xii). Toronto: Pearson Custom Publishing.

Rosina Agyepong, Ph.D., OCT.

is a professor at Centennial College in the Department of English and Liberal Studies, School of Advancement. She teaches Global Citizenship: From Social Analysis to Social Action (GNED 500) and Ethics, Technology, and the Environment (GNED 212). Her research interests include social justice education, feminist studies, spirituality, and alternative approaches to learning.

Philip Alailabo, M.A.

is a political scientist and a professor in the Department of General Education and Liberal Studies at Centennial College. Over the years, he has developed and taught courses in African Politics and Developmental Studies at U.S. universities. Currently, he teaches Global Citizenship: From Social Analysis to Social Action (GNED 500) and Canadian Workplace Experience (GNED 219), which he co-developed in 2011.

Paula Anderton, M.A.

teaches General Education courses at Centennial College, including Global Citizenship and Equity. Her professional background as a magazine journalist and editor included articles on gender bias in medicine and emerging green technologies, and her graduate studies focused on the Victorian "crisis of faith" after Darwin. She is currently developing a multimedia, reflective journal website.

Selom Chapman-Nyaho, M.A.

is an instructor at Centennial College and a Ph.D. candidate in Sociology at York University. A criminologist, his research and publication includes work on youth, risk, and regulation. Selom has taught courses on racism and the law at York University, the University of Ontario Institute of Technology (UOIT), and Lakehead University and worked previously as a Restorative Justice caseworker in Halifax, Nova Scotia.

Agnes Gajewski, Ph.D.

holds a doctorate in education from the University of Toronto. She is a professor at Centennial College teaching on subjects of equity and diversity and psychology. Her research interests include ethics, inclusive education, and disability.

Moreen Jones Weekes, M.A.

is a professor at Centennial College. She teaches several courses including Global Citizenship: From Social Analysis to Social Action (GNED 500). She has also participated in comprehensive revisions to GNED 500 content and assessments since the course launched in 2007. Her educational background is varied and includes Conflict Analysis and Management—Social/Restorative Justice, Mediation, and Arbitration—at an international level.

Athanasios Tom Kokkinias, M.A.

teaches at Centennial College where he has helped develop the college's Global Citizenship course. Tom holds a Master's in Philosophy of Education from OISE/UT, an honours B.A. in Political Science from the University of Toronto, a certificate in Economic Analysis from Ryerson University, and a Teacher of Adults certificate from Centennial College.

Sabrina Malik, M.A.

has been involved in social activism including researching barriers faced by internationally educated professionals in Ontario, increasing access to postsecondary education for marginalized student populations, and building empathy. She now teaches courses in the social sciences and humanities and is actively involved in the creative arts community in Toronto.

Meera Mather, Ed.D.

is the Chair of the General Education and Liberal Studies Department. She oversees the curriculum and delivery of the GNED 500 course as well as various general education and liberal studies courses including the fully online GNED 500 course. In addition to being involved in the GNED 500 comprehensive review and ongoing modifications, Meera has initiated and led the first and current iteration of the GNED 500 textbook.

Cara Naiman, B.A.

is an instructor of Global Citizenship: From Social Analysis to Social Action (GNED 500) at Centennial College. Cara has several years of experience in international and community development working in China, Sri Lanka, western Canada, and Toronto. She has received training in facilitation, learner-centred teaching, and popular education, among other subjects. Currently, Cara is studying toward becoming a psychotherapist.

Jared Purdy, M.A.

has been faculty at Centennial College since 2008. The main subjects that he teaches are global citizenship, ethics, political economy, and propaganda studies. His other areas of interest are Aboriginal history and politics, both nationally and internationally, systemic discrimination, identity politics, and development and conflict issues in Africa.

Chet Singh, Ph.D.

is an educator, social activist, and poet. He has developed and implemented human rights policies and curriculum transformation projects for colleges, universities, and other public sector organizations. He was a founding member and Artistic Director of Canada's Dub Poets Collective and served as a board member for the Ontario Arts Council.

Alia Somani, Ph.D.

completed her doctorate in the English department at the University of Western Ontario. Her areas of interest include postcolonial literature and theory, trauma studies, and Canadian literature and culture. She is currently an instructor at Centennial College where she teaches Global Citizenship and English.

CONTENTS

Chapter 1

Global Citizenship: From Theory to Application

Philip Alailabo, Moreen Jones Weekes, Athanasios Tom Kokkinias, and Cara Naiman

LEARNING OUTCOMES

LO-1 Compare various meanings of global citizenship

LO-2 Examine the historical and theoretical overview of global citizenship

LO-3 Identify reasons for studying global citizenship: big picture thinking, interpersonal skills, reflective practice

As the world becomes smaller, our role in it—and the role of our students—must become larger. To this end, we will strive to become an internationally recognized leader in education that places a strong emphasis on global citizenship, social justice, and equity.

—Centennial College's Book of Commitments*

INTRODUCTION

global citizenship
A concept based on social justice principles and practices that seeks to build global interconnectedness and shared economic, environmental, and social responsibility.

Global citizenship is a foundational concept that is especially relevant in our diverse society. At its most basic, global citizenship is the idea that we belong to a shared community of fellow human beings. That is to say, we are not only citizens of a nation, we are also citizens of the world. This concept is particularly relevant today as individuals, communities, and nations become increasingly interconnected. This course is designed to demonstrate Centennial College's commitment to the ideals of global citizenship, social justice, and equity, which are central to the GNED 500 course—*Global Citizenship: From Social Analysis to Social Action.*

---

* Centennial College, Strategic plan 2009-14: Our book of commitments, 2009. Found at: http://intranet .centennialcollege.ca/corporate/stratplan2009-14

IndianSummer/Shutterstock

What does it feel like when we look at a map of the world with south at the top?

Centennial College has set itself apart from other postsecondary institutions by prioritizing not only the importance of skills and vocational training, but also the essential need for a broader **cultural awareness**[1] and global perspectives in today's workplaces. This course expands on five core concepts that often have significant impact on who we are, our perception of the world and how, in the words of Mahatma Gandhi, we can "be the change we hope to see in the world" through social action. These topics are: social analysis, media literacy, identity and values, equality and equity, and social action. These are concepts that have become instrumental in reminding us of our responsibilities to think and act both locally and globally.

While there are some new concepts and facts to learn in this course, the main focus is on developing your **critical thinking** skills through the lens of global citizenship. Everyone comes with their own opinions and perspectives about the world. This global citizenship class aims to have you challenge your views and look at our world with an analytic eye. That means questioning all aspects of the world around us. For example, when we look at a map of the world, we always see north at the top and south at the bottom. Why is that? Who began this convention and what purpose does it serve? What does it feel like when we look at a map of the world with south at the top?

There are many other ways we can look at our world differently as well as critically. For example, we often take for granted some aspects of contemporary life such as political structures, economic systems, and technological advancements. Challenging these constructs can provide a window to understanding how the

cultural awareness
An ability to interact effectively with people of different cultures and socio-economic backgrounds.

critical thinking
The mental process of actively and skillfully conceptualizing, applying, analyzing, synthesizing, and evaluating information to reach an answer or conclusion.

---

1. Cultural awareness usually involves a deliberate action to learn and understand other cultures or ethnic groups. To enhance one's own views and perspectives about those cultures to facilitate interaction and human development.

world is shaped and determining how to move forward. The content of this course is designed to be used to examine the past and imagine a future in which we can all become engaged on a personal, community, and global scale.

GLOBAL CITIZENSHIP—WHAT DOES IT MEAN?

> Global citizenship requires action, not explanation, to manifest. It is something that must be done not described. (Nekvapil, n.d., p. 2)

The United Nations Academic Impact (UNAI) initiative defines global citizenship as "an umbrella term for the social, political, environmental, or economic actions of globally-minded individuals and communities on a worldwide scale" (UNAI, n.d.). This definition, however, is broad. One might ask what it means to be "globally minded" or how all of these actions are linked to one another. What impact could this interconnectivity have on us as individuals? Furthermore, what roles do civic rights and responsibilities have in the global context?

There is no standard definition of global citizenship, but scholars and writers generally agree on common topics that fit under the umbrella term of global citizenship. Such topics include:

- economic fairness
- equitable distribution of resources
- education
- poverty alleviation
- cultural identity
- the environment
- human rights
- health
- gender equality
- **globalization**
- **social entrepreneurship**[2]
- social justice
- sustainable economic development, and

globalization
The increasing integration of world economies, trade, products, ideas, norms, and cultures in ways that affect all humanity as members of the global community.

social entrepreneurship
A commerce model that combines the principles of business with the objectives of social action or charity.

---

2. There are two models of social entrepreneurship:
    1. "Non-profit with earned income strategies: a social enterprise performing hybrid social and commercial entrepreneurial activity to achieve self-sufficiency. In this scenario, a social entrepreneur operates an organization that is both social and commercial; revenues and profits generated are used only to further improve the delivery of social values.
    2. For-profit with mission-driven strategies: a social-purpose business performing social and commercial entrepreneurial activities simultaneously to achieve sustainability. In this scenario, a social entrepreneur operates an organization that is both social and commercial; the organization is financially independent and the founders and investors can benefit from personal monetary gain" (Abu-Saifan, 2012).

- corporate responsibility towards one another as global citizens (United Nations, n.d.).

Taking action is another aspect of the definition. We can look at a global citizen as someone who believes in and identifies with being part of an emerging global community where he/she is an active participant in shaping its values, culture, norms, and practices (Israel, 2013). In this description, there is an innate understanding of one's role in building this community through individual and collective approaches to achieve the principal objectives outlined above.[3]

To further understand this definition, one has to examine it from the following two lenses; the existence of an emerging global community and the fact that the community has a collection of values, norms, and practices it identifies with (Israel, 2013).

Human beings have the tendency to connect with those who have the same needs or share the same values, beliefs, and ideas. For example, if a Canadian travels to Australia, often one of their first actions is to seek fellow Canadians that are likely to share the same national values or social experiences or seek Australians of like faith. The need to identify often grows into communities with shared values and common objectives that aim to preserve and foster those values through governance and leadership structures (Israel, 2013).

The creation of this community further enables the spreading of technology across the globe that helps individuals to build a sense of belonging. The ability to communicate seamlessly strengthens our ties and connects us to the rest of the world. This is possible through the use of the Internet, our increasing individual and collective global economic activities, and through our feelings of empathy at the sight of inequality in our world (Israel, 2013). However, technology continues to be available in limited ways to the majority of the world's population (National Geographic, 2009).

Global citizenship allows an individual to see the interdependence of nations and the interconnectivity of human activity that provides the impetus to advocate for the disempowered peoples of our world. It means to speak against injustices around the world and to be aware of the process of policy formation, to understand people of diverse cultures, and to learn from interactions with them. It also includes the acquisition of an acute sense of empathy that informs social action not only in one's immediate environment but around the world (Israel, 2013).

When compared, all of these definitions point to two fundamental perspectives. The first is that global citizenship is a way of imagining ourselves in relation to others in the world. One could describe it as a global state of mind. The second is

---

3. There are many people and organizations in the developing world that have difficulty with the liberal, do-gooder notion of global citizenship. This idea is reflected in Ivan Illich's speech, "To Hell with Good Intentions" (Illich, 1968).

that it is a way of acting that carries ethical implications. While it may be difficult to observe people's attitudes, it is easy to see examples of people acting as global citizens.

On January 12, 2010, for example, Haiti suffered a catastrophic earthquake of 7.0-magnitude that affected thousands of people with thousands more losing their lives (CBC, 2011). In the wake of this human tragedy, many countries and individuals around the world responded to appeals for humanitarian aid. Not only were donations collected, but people also travelled from all over to join Haitians in their rescue, recovery, and rebuilding efforts. Yet, here we can also see an example of how the intentions of global citizenship are not enough as billions of aid dollars poured in, but little got done (BBC News, 2013).

The values and ideals of global citizenship can also come under criticism. Is it enough to have good intentions? According to a speech given by Ivan Illich to a group of American youth about to embark on an aid project in Mexico, the answer is no (Illich, 1968). Illich prompts us to ask a number of critical questions: Who decides what help is needed and how it is given? What is the impact of a dominant country exporting its aid workers to the non-dominant community? Why is the aid being sent?

To understand the complex role of global citizenship in today's world, we must begin by examining the historical roots of the concept.

HISTORICAL AND THEORETICAL OVERVIEW OF GLOBAL CITIZENSHIP

In the world of our early human ancestors, social circles were small and knowledge was either passed down from elders or experienced first-hand. Compare that to society today where we live in communities of millions, have access to knowledge far beyond the lore of our elders, and contact with the rest of the world is literally in our hands. This places us all in a powerful position. We can enrich the world with good ideas or we can spread harm (Appiah, 2006).

> Each person you know about and can affect is someone to whom you have responsibilities: to say this is just to affirm the very idea of morality. The challenge, then, is to take minds and hearts formed over the long millennia of living in local troops and equip them with ideas and institutions that will allow us to live together as the global tribe we have become. (Appiah, 2006)*

The concept of global citizenship can provide the modern world with much-needed critical and ethical perspectives on big-picture topics such as social justice

---

*   Anthony Appiah, *Cosmopolitanism: ethics in a world of strangers*. New York: W.W. Norton & Co., 2006.

and equity for all people. Furthermore, global citizens seek to ensure the well-being of animals, plants, and ecosystems on which all life depends. The study of global citizenship provides us with multiple perspectives that can guide us toward understanding and experiencing the world more clearly and in new ways. It offers us diverse and critical considerations of many of today's important issues. Above all else, it provides us with tools to be critically aware of our own biases and agendas. As global citizens, we must consider our impact on communities that we are not members of, but ones that we have, nonetheless, invited ourselves to be a part of.

Although the term global citizenship is relatively new, it is part of a long tradition of something known as cosmopolitanism. The word **cosmopolitan** means "world citizen." Citizenship usually refers to membership and participation in a specific community, which has specific legal rights and duties. For example, a Canadian citizen has to obey the law and pay taxes. In return, Canadians have freedom of conscience, speech, and religion, and the right to vote, among other things (Citizenship and Immigration Canada, 2012).

But those considered to be the early thinkers of global citizenship questioned this notion that citizenship should be tied to one location. The ancient philosopher Diogenes was one of the first documented global citizens. Much like an artist or musician today, Diogenes was notorious for challenging the conventions of Greek society in numerous and often outrageous ways. For example, he lived in a large clay pot in the marketplace of ancient Athens. When asked where he came from, he replied, "I am a citizen of the world."

Philosophers throughout the ancient world had many shared ideas about what global citizenship should involve. Examples of their thinking included treating all people as fellow citizens and having one law common to the entire planet rather than a different set of laws in each country. These early philosophers were united by a belief in the shared humanity of all persons.

The spirit of global citizenship has emerged in many parts of the world over the centuries. Much of South and Southeast Asia was united by language and the activities of wandering traders, writers, religious figures, and adventurers. There was a similar period of cultural flourishing under the Islamic Abbasid Dynasty, which stretched from the Middle East and Persia to North Africa and Spain. Philosophy, science, mathematics, and literature thrived in a cosmopolitan environment that mixed languages, religions, and ethnicities. Global citizenship was also experienced in the great multicultural centres of the Ottoman Empire, including Istanbul, Aleppo, and Baghdad, where people from all faiths and ethnicities lived and worked together (Riedler, 2008; Zubaida, 2010). Chinese philosophers dreamed of a utopian world where national boundaries would be eliminated along with race, class, and gender inequality (Tay, 2010).

Two hundred years ago, a German philosopher expressed the idea that all human beings have an "intrinsic worth" and a "dignity" that must be respected

cosmopolitan
Belonging to all the world; not limited to just one part of the world. To be free from local, provincial, or national ideas, prejudices, or attachments.

(Kant, 1981). Because of this, we cannot treat others simply as a means to our own goals. In everyday language, we must not "use" people. We must therefore take into consideration the good of others when we act (Kant, 1981). This requires us to ask ourselves whether our individual actions could be something that everyone else could also do. Along these lines, others have argued that it should be illegal to treat the citizens of another country in a way that would be forbidden in one's own. In other words, we should not make exceptions for the behaviour of our nation in relation to other nations.

This is one of the most influential arguments in global citizenship studies today. It directly reflects on how imperialistic countries have abused other nations, treating their inhabitants as though they "counted for nothing," causing slavery, famine, and war (Kant, 1983). These endless wars have continued to plague the human species, making it impossible for us to live up to our true potential. Many theorists of global citizenship have hoped that we would one day come to our senses and develop international laws that would be respected around the world.

> Because a…community widely prevails among the Earth's people, a transgression of rights in one place in the world is felt everywhere. (Kant, 1983, p. 119)

WHY STUDY GLOBAL CITIZENSHIP?

With rapid globalization, growing nation-state interdependency and ready access to the Internet, events in other parts of the world have proven to have had profound impacts on the lives of youth and adults. One of these impacts is increased competition for jobs. We will show how studying the concepts of global citizenship helps to develop three important skills that allow candidates to stand out.

First, it builds the practice of big-picture thinking. This is needed to contextualize world events and to understand implications that may motivate political and social action through positive engagement (Oxfam, n.d.).

Second, it develops **interpersonal skills**: the intangible qualities outside of a person's technical expertise that make up one's unique approach to interacting in the world. These qualities include abilities such as problem solving, cultural awareness, and analytic thinking. Global citizenship incorporates these skills through analyzing how people are affected by global issues and understanding the connections between ideas.

Third, it dispels stereotypes through learning about marginalized peoples around the world. It provides the opportunity to challenge age-old assumptions and to assess one's personal heritage through self-reflective practices.

interpersonal skills
Personal attributes that enable someone to interact effectively and harmoniously with other people.

Big-Picture Thinking

Showing an interest in global issues sends a clear message to an employer that an employee is not just a good healthcare worker, business administrator, or engineer, but that the employee is also aware of their community and the interconnectivity of global issues. Globally thinking employees can contribute much more to the workplace by establishing themselves as big-picture thinkers. No work environment operates in isolation, so an understanding of world issues is a strong indicator of one's ability to analyze and solve problems as well as appreciate the complexity of the workplace.

Currently, employers are looking to hire those who come with skill sets that are more than their vocation of study. Employees must be familiar with policies and processes that reflect the interconnectivity of global skills. For example, what are the policies that govern trade, barriers to trade, and the downside to globalization? How do these policies affect our local economy and economies abroad? How might they affect your own studies and future career?

A phrase that has become popular in recent political discourse is "the race to the bottom."[4] It represents a global movement toward strengthening market economies at the expense of disempowered workers and impoverished communities. In an article "Exploring the Global Race to the Bottom," the writer discusses multinational corporations (MNCs) and their involvement in the imbalance of power between the economies of developed and underdeveloped countries (Ashby, Goldstein, & Van Dort, n.d.). While MNCs understand the nuances of becoming developed, they seize the opportunity to use their power to force developing nations to get on board or threaten to take their business elsewhere. As writers Chan and Ross (2003) conclude: "MNCs possess the power for one underlying reason: they are portable."

By examining concepts such as the race to the bottom and the role of MNCs in the global economy, we start to look beyond the day-to-day nature of our jobs and careers. It enables us to consider the big picture of how our individual work is linked to a global system. In addition, we cannot rely on the viewpoint of a privileged few who are in power, since this will not provide a sufficiently diverse perspective to paint an accurate picture of world issues. We must participate in the dialogue and bring our perspectives to the discourse.

---

4. This can also be looked at as a corporate agenda to pay workers as little as possible so that those in power can earn more profit. These corporations will go to countries that will facilitate the necessary changes to their labour laws, environmental protection acts, etc. These massive, multinational corporations set the world labour standards that others will follow.

Khalil Bendib, www.bendib.com

Interpersonal Skills

The Conference Board of Canada (2000) produced a range of employability skills necessary for success and advancement in employment. These skills are: fundamental skills, personal management skills, and teamwork skills. These are then expanded to provide detail on the skills needed to enter, stay in, and progress in the world of work (The Conference Board of Canada, 2000).

Interpersonal or life skills are critical in the workplace. These are not vocational skills—specific to a discipline of study—but are necessary to understand, communicate, and interact with co-workers as well as clients to whom the organization provides a product or service. To be successful, an employee must be comfortable dealing with co-workers and international clients as well as work effectively within groups.

Canadian employers want employees who can resolve conflict at the earliest stage, think critically, communicate effectively, and engage successfully with a wide range of people of various ethnicities. For this reason, it is important for students to explore their conflict style with a view to understanding if their approach is to be an avoider, a competer, a compromiser, a collaborator, or an accommodator (Thomas & Kilmann, 1977).

Students must understand how to resolve conflicts in an open, transparent, and non-confrontational manner. They must learn how to interact with co-workers and clients from a wide multicultural spectrum.

Development of strong interpersonal skills signals a higher rate of success, not just in one's professional life, but also in personal relationships. *Global Citizenship* provides the opportunity to hone interpersonal skills through group work that incorporates both critical thinking and social analysis.

Reflective Practice

Canadian employers also want to hire people who can make meaningful contributions to the workplace. These contributions can be made on a personal, team, organizational, or industry level. In order to contribute, however, one must first look at how things are currently done. By observing an existing practice or procedure, one can notice what works well and where improvements might be needed. These observations may lead to the development of new ideas that can be tested and then further assessed. Refer to Figure 1.1.

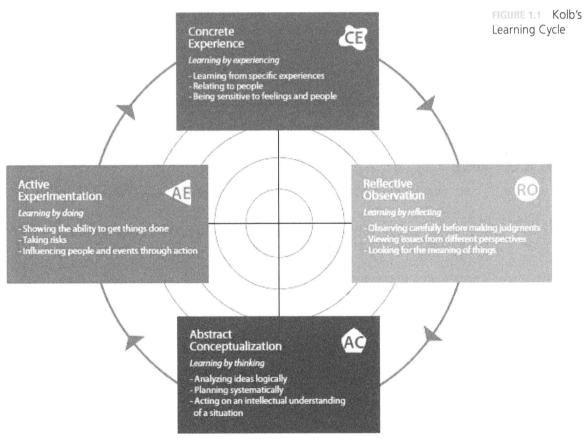

FIGURE 1.1　Kolb's Learning Cycle

Concrete Experience CE
Learning by experiencing
- Learning from specific experiences
- Relating to people
- Being sensitive to feelings and people

Active Experimentation AE
Learning by doing
- Showing the ability to get things done
- Taking risks
- Influencing people and events through action

Reflective Observation RO
Learning by reflecting
- Observing carefully before making judgments
- Viewing issues from different perspectives
- Looking for the meaning of things

Abstract Conceptualization AC
Learning by thinking
- Analyzing ideas logically
- Planning systematically
- Acting on an intellectual understanding of a situation

Source: Courtesy Hay Group Inc, 2007. 399 Boylston St, Boston MA. www.haygroup.com/leadershipandtalentondemand

reflective practice

The capacity to reflect on action so as to engage in a process of continuous learning.

The ability to incorporate **reflective practice** leads very directly to creating innovation in the workplace. Google, a company renowned for its innovation, applies its eight pillars of innovation principles in its corporate culture to ensure that all employees are looking at ways of doing things differently through a reflective practice lens. Six of these pillars are summarized here:

- The most effective way to tackle a major initiative is to take small steps toward the larger goal.
- Innovating means making continuous improvements. You won't get it 100% right the first time.
- Ensure that all ideas are brought forward—every idea is important.
- Sharing all ideas and data with the whole team is the best way to encourage innovation.
- Be open to all possibilities and then test them to see if they're feasible.
- It's OK to fail if you've learned something from that failure (Google, n.d.).*

Approaches such as this encourage employees to explore existing strategies, policies, or procedures, experiment with making changes to them, learn from the experience and apply that learning to improving the idea. The result is a workplace made up of engaged personnel in an environment that is responsive to the increasingly fast-paced global shifts.

CONCLUSION

Throughout this text we will explore core units that are vital to the role of being a global citizen and will further develop the analytical and critical thinking skills needed to survive in the competitive market of employment. We will examine why it is important to conduct a social analysis of the injustices that plague society to gain a better understanding of the root of the problem. In Chapter 2, we will further explore the concept of global citizenship and its relationship to both national citizenship and globalization. Chapter 3 discusses the injurious harm that occurs when individual behaviours that are based on ideologies are central to contributing to injustices. Chapter 4 further develops the analysis that must occur to get to the root of the problem and presents frameworks for examining issues and ideologies. In Chapter 5, the influence of the media is explored and how this influence contributes to the way we interpret our surroundings. The complexities inherent in identity, whether social or personal, are explored in Chapter 6. The intention here is to use self-reflection and analysis to understand how identities created by society play a major role in constructing identities on a personal level. Chapter 7 engages the historical context of injustices through the struggles of indigenous peoples

---

*   Adapted from: Susan Wojcicki, "The Eight Pillars of Innovation," *Think Quarterly*, Google. Found at: http://www.google.ca/think/articles/8-pillars-of-innovation.html

and how the contribution of this oppressed group has resulted in global change. With the combined knowledge of the previous chapters, personal and professional responsibilities are explored in Chapter 8 through the discussion of the possibilities for social action.

Centennial College believes that thinking and acting as global citizens results in increased engagement in social issues. This, in turn, leads to more positive and stronger communities. You are invited to enter this course with your opinions, assumptions, biases, and beliefs. The hope is that throughout the term, you will remain open to challenging yourself and consider new possibilities toward making change. This process requires an openness to differences and diversity that can be difficult at times, but, like all struggles, has the potential to produce rich and lasting results.

> No matter how complex global challenges may seem, we must remember that it is we ourselves who have given rise to them. It is therefore impossible that they are beyond our power as human beings to resolve. Returning to our humanity, reforming and opening up the inner capacities of our lives, can enable reform and empowerment on a global scale. (Ikeda, n.d.)*

CRITICAL THINKING QUESTIONS

1. What does being a global citizen mean to you?
2. What are some different ways of thinking of global citizenship?
3. How can global citizenship have positive effects on the world and how can its effects be negative?
4. How might issues of global citizenship and globalization affect your own studies and future career?

REFERENCES

Abu-Saifan, S. (2012). Social entrepreneurship: Definition and boundaries. *Technology Innovation Management Review*. Retrieved from http://timreview.ca/article/523

Appiah, A. (2006). *Cosmopolitanism: Ethics in a world of strangers.* Retrieved from http://books.google.ca/books/

Ashby, M., Goldstein, Z., & Van Dort, C. (n.d.). *A "race to the bottom:" The adverse effects of globalization on environmental standard.* Retrieved from http://www.globalchange.umich.edu/globalchange2/current/workspace/Sect007/s7g3/RTB%20effect.htm

---

*   Daisaku Ikeda, Words of wisdom: Buddhist inspiration for daily living. Found at: http://www.ikedaquotes.org/global-citizenship/globalcitizenship454

BBC News. (2013, January 12). Haiti President Martelly criticises aid on quake anniversary. *BBC News Latin America and Caribbean*. Retrieved from http://www.bbc.com/news/world-latin-america-21001060

CBC. (2011, January 5). Special report, Haiti Earthquake: A look back, 2 years after disaster crippled Caribbean country. *CBC World News*. Retrieved from http://www.cbc.ca/news/world/special-report-haiti-earthquake-1.1137266

Centennial College. (2008). *Global citizenship: From social analysis to social action* (2nd ed.). Toronto: Pearson Custom Publishing.

Centennial College. (2009). *Strategic plan 2009–14: Our book of commitments.* Retrieved from http://intranet.centennialcollege.ca/corporate/stratplan2009-14

Chan, A., & Ross, R. (2003). Racing to the bottom: International trade without a social clause. *Third World Quarterly, 24*(6), 1011–028.

Citizenship and Immigration Canada. (2012). *Rights and responsibilities of citizenship.* Retrieved from http://www.cic.gc.ca/english/resources/publications/discover/section-04.asp

Conference Board of Canada. (2000). *Employability skills 2000+.* Retrieved from http://www.conferenceboard.ca/topics/education/learning-tools/employability-skills.aspx

Google. (n.d.). Retrieved from http://www.google.ca/think/articles/8-pillars-of-innovation.html

Ikeda, D. (n.d.). Words of wisdom: Buddhist inspiration for daily living. Retrieved from http://www.ikedaquotes.org/global-citizenship/globalcitizenship454

Illich, I. (1968). To hell with good intentions. Retrieved from http://www.swaraj.org/illich_hell.htm

Israel, R. (2013). What does it mean to be a global citizen? Retrieved from http://www.opendemocracy.net/ourkingdom/ron-israel/what-does-it-mean-to-be-global-citizen

Kant, I. (1981). *Grounding for the metaphysics of morals* (J. W. Ellington, Trans.). Indianapolis: Hackett.

Kant, I. (1983). To perpetual peace: A philosophical quest. In *Perpetual peace and other essays* (T. Humphrey, Trans.). Indianapolis: Hackett.

National Geographic. (2009). *The world of 7 billion* [Visual data representation]. Retrieved from http://ngm.nationalgeographic.com/2011/03/age-of-man/map-interactive

Nekvapil, E. (n.d.). Why global citizenship? Retrieved from http://www.abolishforeignness.org/blog/wp-content/uploads/2011/04/Why-Global-Citizenship.pdf

Oxfam. (n.d.). *Why is education for global citizenship essential in the 21st century?* Retrieved from http://www.oxfam.org.uk/~/media/Files/Education/Global%20Citizenship/education_for_global_citizenship_a_guide_for_schools.ashx"tizenship_a_guide_for_schools.ashx

Riedler, F. (2008, Autumn). Rediscovering Istanbul's cosmopolitan past. *ISIM [International Institute for the Study of Islam in the Modern World] Review, 22,* 8–9.

Tay, W. L. (2010). Kang Youwei, the Martin Luther of Confucianism and his vision of Confucian modernity and nation. In Secularization, religion and the state. University of Tokyo Center for Philosophy, *Booklet 17,* 102–03.

Thomas, K., & Kilmann, R. (1977). Developing a forced-choice measure of conflict-handling behavior: The "MODE" instrument. Retrieved from http://epm.sagepub.com/content/37/2/309.short

UNAI (United Nations Academic Impact). (n.d.). Retrieved from http://unai-globalcitizenship.org/global-citizenship

United Nations, *Charter of the United Nations*, 1945; *Universal Declaration of Human Rights*, 1948. See also *Abolition of Forced Labour*, 1957; *International Convention on the Elimination of All Forms of Racial Discrimination*, 1965; and *Convention against Torture and Other Cruel, Inhuman or Degrading Treatment or Punishment*, 1984.

Zubaida, S. (2010, July 20). Cosmopolitan citizenship in the Middle East. Retrieved from http://www.opendemocracy.net/sami-zubaida/cosmopolitan-citizenship-in-middle-east

Chapter 2

A Critical Analysis of Global Citizenship

Philip Alailabo, Moreen Jones Weekes, Athanasios Tom Kokkinias, and Cara Naiman

LEARNING OUTCOMES

LO-1 Consider the need for global citizenship through examples

LO-2 Distinguish between global citizenship and national citizenship

LO-3 Critically analyze global citizenship and globalization from an economic, political, and socio-cultural perspective

INTRODUCTION

Having looked at the history of global citizenship, we can now examine some of the roles this concept has in the world with the aim of distinguishing it from national citizenship and globalization. While reading this chapter, students are invited to consider these questions: How can we move toward developing a better understanding of the idea of global citizenship? Is global citizenship the responsibility of one person or does it take a village of people working together to achieve a common goal? Will lived experience change what it means to be a global citizen over time?

By exploring the concept of global citizenship from its historical roots to the current day, the thread that binds the units in this chapter becomes clear. As we move from social analysis to identity construction, media literacy, and social action, the evidence of inequality and inequity comes to light. In addition, gaining a deeper understanding of our roles as members of society through the art of self-reflection may serve to solidify the need for becoming global citizens.

EXAMPLES OF THE NEED FOR GLOBAL CITIZENSHIP

GNED 500 provides the opportunity to engage people not only on an individual and community level, but also on a global scale. While the historical contexts studied in Chapter 1 are important to understanding where we came from, experiences in a contemporary context illustrate the need to continue our awareness and think critically about the global impact of our actions. Consider these contemporary examples of social injustices.

In September 2013, students at a Canadian university engaged in chanting about non-consensual sex with underage young girls. In August 2013, during an event welcoming newcomers to another university, students shouted chants promoting sexual misconduct towards virgins. Through these acts, educational institutions are viewed as condoning and normalizing sexual assault as part of an educational culture. This then contributes to the implied acceptance of sexual harassment, not only in postsecondary institutions but potentially also in public schools (Taber, 2013).

In August 2012, a Toronto restaurant was accused of racism for naming new burgers on their menu "Half-Breed" and "Dirty Drunken Half-Breed" (Poisson, 2012). Likewise, in November 2012, Victoria's Secret showed a scantily clad model wearing an Indian headdress (Bergin, 2012). These acts of **cultural appropriation**[1] (in the case of Victoria's Secret) and racism (in the case of the restaurant) show a total disrespect of indigenous practices and culture and serve to further undermine and marginalize a disenfranchised group. Interestingly, in response to accusations of racism and cultural appropriation, both companies claimed ignorance. "We didn't know," they said. But in today's culturally diverse climate, people responded by saying that ignorance is unacceptable: you must be aware of these issues; you must know.

> **cultural appropriation**
> The act of taking on or making use of a non-dominant culture without the authority or right to do so.

That is why a course on global citizenship is important. Even to run a business, you need to be aware of the world around you—a world made up of multiple cultures, histories, and communities. We must consider the real and observable consequences of allowing the violation of human rights of others. In the words of Desmond Tutu (1976), "If you are neutral in situations of injustice, you have chosen the side of the oppressor. If an elephant has its foot on the tail of a mouse and you say that you are neutral, the mouse will not appreciate your neutrality." In school, at work and in our communities, we bear witness to injustices. Being able to speak out and take action is a responsibility we all share as global citizens.

HOW IS GLOBAL CITIZENSHIP DIFFERENT FROM NATIONAL CITIZENSHIP?

When looking at the differences between global citizenship and national citizenship, it is useful to note that the two are not mutually exclusive but rather complementary. National citizenship[2] is usually seen in terms of rights, privileges, and

---

1. Cultural appropriation is explored further in Chapter 5, Media Literacy.
2. National citizenship can also carry implicit racism, religious discrimination, and xenophobia (fear of strangers or the unknown) as has been characterized in Eastern Europe, Rwanda, Uganda, Swaziland, Germany, Japan, and other countries.

responsibilities with loyalty to a state whose powers in democratic structures are sustained by its citizens through periodic elections. It is obtained through birth or naturalization and includes shared national values along with obligations such as paying taxes and following the law. There is an expectation of shared responsibility where citizens fulfill their obligations to the state and the state, in turn, fulfills its duty by providing such things as protection, education, health care, and access to jobs and resources.

On the other hand, global citizenship is undertaken willingly. As Schattle (2008) suggests, it implies "ways of thinking and living within multiple cross-cutting communities—cities, regions, states, nations and international collectives…." It is demonstrated through association rather than legality as with national citizenship (Lagos, n.d.).

Identifying with the concepts of global citizenship does not undermine the value of our national citizenship. It simply means that we assume greater responsibilities either at the local or international level through engagement in community or global issues. However, our notion of global citizenship may also challenge our understanding of identity and ways in which we contribute to inequity. It can prompt us to question what we think is right.

One way that these global responsibilities are upheld and maintained is through the United Nations *Universal Declaration of Human Rights*. The United Nations, founded in 1945 following the devastation of World War II, provides an example of trying to legitimize the ideals embodied by the values of global citizenship (Schattle, 2008). In 1948, it negotiated the *Universal Declaration of Human Rights*. Here are some of the key principles:

- Everyone has the right to life, liberty, and security of person (Article 3).
- No one shall be held in slavery or servitude; slavery and the slave trade shall be prohibited in all their forms (Article 4).
- No one shall be subjected to torture or to cruel, inhuman, or degrading treatment or punishment (Article 5).*

Today, 193 countries are members of the United Nations and many other international conventions have been negotiated. Although the United Nations cannot force nations to comply with the treaties they have signed, it exercises moral power through its meetings, complaints process, and annual progress reports.[3]

---

* United Nations, *Charter of the United Nations, 1945; Universal Declaration of Human Rights, 1948*.

3. While it is true that the General Assembly of the United Nations enjoys this broad international membership, some rightly point out that in practice, the UN is far from the ideal of a pluralistic and broadly egalitarian force on the global level. For instance, many criticize the UN for a number of high profile policy failures on the world arena, including the well-known fact that the tiny minority of states comprising the veto-holding Security Council, most notably the U.S., have dominant control over the various bodies, processes, and mandates of the organization. As one author put it, "The widely held impression among UN insiders today is that the United Nations remains largely a U.S.-controlled organization just as it has been for the last half-century. The difference now is that its control is seen as more compelling than ever before" (Puchala, 2005, p. 574).

Another difference between national and global citizenship is the fact that global citizenship has no privileges, duties, or recognizable rights that are firmly associated with the concept that would grant legal status and power on the global citizen (Lagos, n.d.). With no recognizable rights or privileges, a global citizen is often motivated by the inner satisfaction gained through engagement and the feeling of accomplishment that comes from making a difference.

However, it should be noted that since the 1970s, there has been some movement toward instituting multinational laws that some philosophers have been proposing for hundreds of years. Many regional bodies consisting of countries across the world have put in place laws, policies, or trade agreements that appear to sanction the concept of global citizenship thereby giving it political and legal recognition. For example, the 15 countries in West Africa under the umbrella of Economic Community of West African States (ECOWAS) have waived visa requirements for their citizens to travel within the region. The objective of this waiver is to facilitate trade and allow citizens to engage in cross-border activities aimed at poverty alleviation and achieving greater economic stability (1979 Protocol).

HOW IS GLOBAL CITIZENSHIP DIFFERENT FROM GLOBALIZATION?

Chapter 1 presented a number of ways to think about global citizenship. Here are two more definitions to consider. The first was established by Centennial College's Institute for Global Citizenship and Equity:

> To be a citizen in the global sense means recognizing that we must all be aware of our use of the world's resources and find ways to live on earth in a sustainable way. When we see others are treated without justice, we know that we are responsible for trying to ensure that people are treated justly and must have equitable opportunities as fellow citizens of this world. We must think critically about what we see, hear and say, and make sure that our actions bring about positive change. (Centennial College, 2008)*

The second definition comes from Oxfam, a major international aid organization. It defines a global citizen as someone who

- is aware of the wider world and has a sense of their own role as a world citizen
- respects and values diversity
- has an understanding of how the world works economically, politically, socially, culturally, technologically and environmentally
- is outraged by social injustice

---

* Centennial College's Institute for Global Citizenship and Equity, 2008.

- participates in and contributes to the community at a range of levels from local to global
- is willing to act to make the world a more sustainable place
- takes responsibility for their actions. (Oxfam, n.d.)*

Through these definitions, we can see that social justice, intercultural tolerance, and environmental concern are embedded into the meaning of global citizenship. Compare these concepts to those described in the following definition of globalization.

> [The] process by which the experience of everyday life, marked by the diffusion of commodities and ideas, is becoming standardized around the world. Factors that have contributed to globalization include increasingly sophisticated communications and transportation technologies and services, mass migration and the movement of peoples, a level of economic activity that has outgrown national markets through industrial combinations and commercial groupings that cross national frontiers, and international agreements that reduce the cost of doing business in foreign countries. Globalization offers huge potential profits to companies and nations but has been complicated by widely differing expectations, standards of living, cultures and values, and legal systems as well as unexpected global cause-and-effect linkages. (Merriam-Webster, n.d.)†

Another way to describe globalization is the increasing integration of world economies, trade, products, ideas, norms, and cultures in ways that affect all humanity as members of the global community (Albrow & King, 1990; Al-Rodhan & Stoudmann, 2006). The discussion of globalization includes the improvements in transportation and the steady rise of telecommunications. These advancements have enabled people to travel long distances quickly and communicate effortlessly through the Internet. This allows us to be informed about events around the world, sometimes within seconds, and enables us to take action to support strategies to bring about change. From another perspective, globalization can be seen by some to have an economic focus.

Globalization carries negative associations for those who see it as a means to establish corporate powers at the expense of the disenfranchised, whereas most people agree that global citizenship is a desirable social goal. Consequently, a multinational corporation might choose to apply the vocabulary of global citizenship in their marketing material to give consumers the impression that they are using ethical, environmentally responsible, and transparently fair business practices. In

---

* Oxfam Education. Reproduced from www.oxfam.org.uk/education/global-citizenship/what-is-global-citizenship with the permission of Oxfam GB, Oxfam House, John Smith Drive, Cowley, Oxford OX4 2JY, UK. Oxfam GB does not necessarily endorse any text or activities that accompany the materials.

† By permission. From Merriam-Webster's Collegiate® Dictionary, 11th Edition ©2014 by Merriam-Webster, Inc. (www.Merriam-Webster.com).

fact, the company might not always operate in such a way, but could argue that, to them, global citizenship means having an impact on people around the world while ignoring any potentially negative results from that impact.

Here are a few examples of major multinational companies' mission statements:

"We save people money so they can live better." (Walmart Stores, n.d.)*

We "create a family of devices and services for individuals and businesses that empower people around the globe at home, at work and on the go, for the activities they value most." (Microsoft, 2013)†

"To refresh the world.... To inspire moments of optimism and happiness.... To create value and make a difference." (Coca-Cola Company, 2013)††

Notice the phrases that capture the spirit of global citizenship's goals such as "live better," "empower people," and "make a difference." Many people feel that these companies do, indeed, live up to these statements. However, as critical thinkers, we might also ask which people are living better and is it at the expense of others? Who is empowered and what might be the consequences of the difference that is being made? Are these statements reflective of genuine concern for humanity or are they part of a corporate agenda to transform ideals into commodities?

The production and distribution of food is an example of how the complexities of global citizenship and globalization can have different effects on people and the environment. In our grocery stores, we can buy food from around the world. This is one way that we are able to connect with other countries and cultures. But there are many layers of economic, political, social, and environmental forces that contribute to the availability of food in our stores. These forces act as links in the web of global citizenship and globalization of which we are all a part.

While there are many distinct aspects to globalization and global citizenship, two concepts can be seen as elements of both: that we all depend on one another and we need to work collaboratively. The first highlights the fact that nations of the world are fundamentally connected to one another—events in one country invariably affect other countries across the globe. The second concept, collaboration, speaks to the fact that some of the most daunting problems facing our world today can only be met with a collective approach involving nations around the world. With the enormity of problems in our world such as human trafficking, global warming,

---

* Walmart.
† Microsoft. (2013). Annual report 2013. Found at: http://www.microsoft.com/investor/reports/ar13/shareholder-letter/index.html
†† Coca-Cola Company. (2013). 2012-2013 Sustainability report. Retrieved from http://www.coca-colacompany.com/our-company/mission-vision-values

wars, hunger, poverty, and much more, it has become virtually impossible for any nation to single-handedly provide solutions to these problems (Clinton, 2010).

This means that sustainable solutions must be borne out of collaborative efforts of nations to effectively tackle these problems. For example, widespread human trafficking across international borders means that countries must work together to strengthen applicable laws, share information and technology while improving policing, prosecution, and enforcement.

ECONOMIC AND POLITICAL EFFECTS

trade sanctions

One or more trade barriers that one country places upon another country as a punitive action.

subsidy

A sum of money granted by the state or a public body to help an industry or business keep the price of a commodity or service low.

banana republic

A small state that is politically unstable as a result of the domination of its economy by a single export controlled by foreign capital.

cash crops

A crop produced for its commercial value rather than for use by the grower.

To begin, the economic aspects of global citizenship and globalization would be concerned with issues such as the cost of growing, processing, packaging, and transporting food around the world. In addition, there are **trade sanctions**, farming **subsidies** and world trade markets that affect the importing and exporting of food. The documentary *Black Gold* gives a stark example of how these international commodities markets succeed in maintaining a comfortable standard of living in industrialized countries at the expense of the lives of the producers. Specifically, *Black Gold* demonstrates how people in developed nations have access to some of the best coffee in the world just around the corner at the local coffee shop while, in Ethiopia, the coffee growers live in poverty. The profits are in the billions of dollars, but almost all of it goes to a handful of multinational companies (Francis, Francis, & Hird, 2006).

Another issue that this film touches on is the phenomenon of the **banana republic**. This is a country or region that relies on **cash crops** rather than a more sustainable approach to farming that applies principles of diversity and self-sufficiency. Cash crops often weaken or destroy the nutrients in the soil because it neglects the natural biodiversity that comes from crop rotation (Bareja, 2010). As well, these crops are often subject to the fluctuations of world commodities markets that are controlled by rich, industrialized countries and their multinational corporations. Although the act of buying a cup of coffee might seem small, it is part of a large, interconnected system and the effects can be felt across the globe.

Canadians buy and sell products to and from the far corners of the globe, potentially affecting the lives of people we do not know and will never meet. But does globalization lead to global citizenship? Global citizenship scholars contend that trade agreements tend to benefit the wealthy and powerful western countries, disadvantaging other parts of the world.[4]

---

4. The World Trade Organization (WTO) endorses agricultural subsidies in industrialized countries, but not for developing nations. Consequently, governments of wealthy countries keep produce prices low and poorer countries are unable to compete in the international market (Food and Agriculture Organization of the United Nations, 2003).

For example, the West entered World Trade Organization agreements with Myanmar, the Democratic Republic of the Congo, and Indonesia, even though these countries were ruled by corrupt dictators or military governments (Pogge, 2008). Canada signed a trade deal in 2011 with the Honduran government, which, following a military coup, had won an election not recognized by most international observers (Council of Canadians, 2011). We sell weapons to these countries, look the other way to the payment of bribes by our businesspeople, and provide safe havens for wealth stolen by their elite. Perhaps even more destructively, the West enters lending arrangements with the rulers of these countries that further indebt their nations and purchase resources that inadequately compensate their people. These undervalued resources—of which we use a disproportionate amount—are a key component of our relative wealth (Pogge, 2008).

Looking once again at food production, we can see that political influences are far reaching. Wealthier countries such as Canada and the United States are able to subsidize their farmers and control tariffs on food produced in other countries so that some countries are able to access these markets but others cannot (Food and Agriculture Organization of the United Nations, 2003). For example, we can see apples from New Zealand in the grocery store, but we do not see bananas from Uganda. Political relationships are constantly in effect behind the scenes.

We can also see the effects of globalization on food production with farmers in India. Dr. Vandana Shiva has been devoting years of research and advocacy to addressing the harmful effect of the practices of large seed producers such as Monsanto. Using technology to create genetically modified and sterile seeds prevents farmers from using their traditional methods of sustainable agriculture such as seed sharing. Monsanto has been using patents to force farmers to pay royalties and, in some cases, go into debt just to get access to seeds that should be a renewable resource available to everyone (Shiva, 2013).

Naturally, different political groups in different parts of the world will have a distinct understanding of the meaning of global citizenship. However, an additional challenge to consider is sustainability. In order to achieve meaningful progress in the areas of education, poverty reduction, peace, the environment, etc., ongoing and consistent efforts are required. As discussed earlier, many governmental systems involve voting that allows for changes in ideological stances at fairly regular intervals. This means that the global interests of one ruling body may be quite different from the next. The result is that initiatives which require long-term solutions are often given lower priority.

In addition, the voters themselves may not give these complex global issues much thought if they are struggling to manage within their own daily realities. Consequently, voters will not voice their concerns and politicians can ignore these problems knowing that it will not affect their re-election. We can see this in the election process. In Canada, political platforms are typically made up of local, short-term fixes such as lowering taxes, creating jobs, and improving infrastructure and education systems. Rarely do we hear election campaigns built on working toward ending world hunger or eliminating global conflicts.

SOCIO-CULTURAL EFFECTS

Within the context of global citizenship, there are a number of complex social and cultural factors (socio-cultural effects) that impact the relationship between the effects of globalization and the ethical responsibilities of global citizenship. Globalization creates economic, political, social, and cultural connections, but power structures within these systems can result in maintaining inequality.

Global citizenship is not without its criticism either. Some individuals have expressed concern that global citizenship requires that we uphold a single, universal standard of right and wrong, and, therefore, such a standard has the potential to be oppressive of other cultures. For example, some cultures are tolerant of homosexuality while others are not. If there are to be international laws applicable to all, who will define the rights and responsibilities? Kwame Anthony Appiah has tried to address this criticism. He conceives of a *rooted cosmopolitanism* that values diversity and embraces local differences.

Appiah reminds us that global citizenship is not exclusively a western ideal. When his father, Joseph Emmanuel, died, he left a message for his children reading, "Remember that you are citizens of the world" (Appiah, 2006). Yet Joseph was profoundly patriotic, fighting for the independence of his native Ghana. Appiah fondly recalls Kumasi, his hometown, with its diverse community of Indians, Iranians, Lebanese, Syrians, Greeks, Hungarians, and English. The multilingual global citizen is just as likely to be rich as poor because people have been migrating in search of better lives for millennia (Appiah, 2006).

Appiah draws attention to the respect for the worth of each individual. The dignity of each human is rooted in our capacity to make rules and choices for ourselves. Respecting human dignity and personal autonomy means respecting diversity, including cultural diversity. Global citizens since the time of Diogenes have believed that there are many acceptable ways to live. It is respect for the dignity and worth of the individual that leads to respect for cultural differences (Appiah, 2008).

However, there must be some constraints placed on cultures, including our own, if the philosophy of global citizenship is to have any meaning. Many argue that every human being is entitled to basic human rights that must be protected. Every human being has certain minimal obligations as well. We are each responsible to ensure, to the extent possible, that all other people on the planet receive their fair share. This is referred to as entitlement to "a dignified human existence" (Appiah, 2008). Global citizenship means that we should do no less for our fellow citizens of the world.

CONCLUSION

We acknowledge that, although there is no single, accepted definition of global citizenship, there are three principles that ancient and contemporary philosophies share. Perhaps the defining principle is the belief that there is an ethical code operating above the laws of nations, producing both rights and responsibilities. The second principle is that of fundamental respect for persons. The third principle is the duty to reform systems that are producing inequalities from which we in the West benefit at the expense of others around the world. This principle requires that we take action rather than just talk about global citizenship.

The question of who is able to participate in global citizenship remains—is this something reserved for those with privilege in the developed parts of the world or is it indeed for everyone? The middle classes in industrialized countries might have the luxury of more leisure time to consider things like gender equality and fair access to resources, however, one could argue that it is, in fact, the less privileged of society who are fighting hardest for a more balanced world to live in. One thing that cannot be argued is that the challenges facing the world are vast and complex. We need input and participation from all corners of society in order to get the best results by including all of these diverse perspectives.

> The differences between people need not act as barriers that wound, harm and drive us apart. Rather, these very differences among cultures and civilizations should be valued as manifestations of the richness of our shared creativity. (Ikeda, n.d.)*

CRITICAL THINKING QUESTIONS

1. How can we develop a better understanding of the idea of global citizenship?

2. What roles do civic rights and responsibilities have in the global context?

3. Is global citizenship the responsibility of one person or does it "take a village" of several people working together to achieve a common goal?

4. Will lived experience change what it means to be a global citizen over time?

5. How is global citizenship different from globalization?

6. Do we need to be global citizens? Why or why not?

7. Is global citizenship something reserved for those with privilege in the developed parts of the world or is it indeed for everyone?

---

* Daisaku Ikeda, Words of wisdom: Buddhist inspiration for daily living. Found at: http://www.ikedaquotes .org/global-citizenship/globalcitizenship454

REFERENCES

Albrow, M., & King, E. (Eds.). (1990). *Globalization, knowledge and society.* London: Sage.

Al-Rodhan, N. R. F., & Stoudmann, G. (2006). Definitions of globalization: A comprehensive overview and a proposed definition. Retrieved from http://www.sustainablehistory.com/articles/definitions-of-globalization.pdf

Appiah, K. (2006). *Cosmopolitanism: Ethics in a world of strangers.* New York: W.W. Norton.

Appiah, K. (2008). Education for global citizenship. *Yearbook of the National Society for the Study of Education, 107*(1), 93.

Bareja, B. (2010). *Crop agriculture review.* Retrieved from http://www.cropsreview.com/cash-crop-farming.html

Bergin, O. (2012, November 12). Victoria's Secret apologises over American Indian outfit in catwalk show. *The Telegraph.* Retrieved from http://fashion.telegraph.co.uk/article/TMG9672704/Victorias-Secret-apologises-over-American-Indian-outfit-in-catwalk-show.html

Centennial College. (2008). *Global citizenship: From social analysis to social action* (2nd ed.). Toronto: Pearson Custom Publishing.

Clinton, B. (2010, February 24). Turning good intentions into positive action. Presentation at University of California, Berkeley. Retrieved on December 19, 2013, from http://www.cic.gc.ca/english/department/media/releases/2013/2013-07-03.asp

Coca-Cola Company. (2013). *2012–2013 Sustainability report.* Retrieved from http://www.coca-colacompany.com/our-company/mission-vision-values

Council of Canadians. (2011, March 9). Canada–Honduras FTA about mining and sweatshop protection, foreign affairs committee hears. Retrieved from http://www.canadians.org/tradeblog/?p=1380.

Food and Agriculture Organization of the United Nations. (2003, September). Subsidies, food imports and tariffs key issues for developing countries. Retrieved from http://www.fao.org/english/newsroom/focus/2003/wto2.htm

Francis, M., Francis, N., & Hird, C. (Producers), & Francis, M., & Francis, N. (Directors). (2006). *Black gold* [Motion picture]. England: Speakit.

Ikeda, D. (n.d.). Words of wisdom: Buddhist inspiration for daily living. Retrieved from http://www.ikedaquotes.org/global-citizenship/globalcitizenship455.html

Lagos, T. G. (n.d.). Global citizenship—Towards a definition. Retrieved from https://depts.washington.edu/gcp/pdf/globalcitizenship.pdf

Merriam-Webster. (n.d.). Definition of globalization. Retrieved from http://www.merriam-webster.com/concise/globalization

Microsoft. (2013). *Annual report 2013.* Retrieved from http://www.microsoft.com/investor/reports/ar13/shareholder-letter/index.html

1979 *Protocol A/P.1/5/79 relating to free movement of persons, residence and establishment,* article 2.

Oxfam Education. (n.d.). What is global citizenship? Retrieved from https://www.oxfam.org.uk/education/global-citizenship/what-is-global-citizenship

Pogge, T. (2008). *World poverty and human rights* (2nd ed., p. 28). Cambridge: Polity Press.

Poisson, J. (2012, August 29). "Half-Breed" and "Dirty Drunken Half-Breed" chucked from burger menu. *The Star*. Retrieved from http://www.thestar.com/news/gta/2012/08/29/halfbreed_and_dirty_drunken_halfbreed_chucked_from_burger_menu.html

Puchala, D. J. (2005). World hegemony and the United Nations. *International Studies Review, 7*(4), 571–84. doi: 10.1111/j.1468-2486.2005.00533.x

Schattle, H. (2008). *The practices of global citizenship*. Lanham, Maryland: Rowman & Littlefield).

Shiva, V. (2013). *The seeds of suicide: How Monsanto destroys farming*. Retrieved from http://www.globalresearch.ca/the-seeds-of-suicide-how-monsanto-destroys-farming

Taber, J. (2013, September 5). Saint Mary's student president says rape chant was "biggest mistake…probably in my life." *The Globe and Mail*. Retrieved from http://www.theglobeandmail.com/news/national/saint-marys-student-president-says-rape-chant-was-biggest-mistake-of-my-life/article14142351/

Tutu, D. (1976). Retrieved from http://www.biographyonline.net/spiritual/desmond-tutu.html

Walmart Stores. (n.d.). Mission statement. Retrieved from http://corporate.walmart.com/our-story/

Chapter 3

Social Analysis for Social Change

Chet Singh

All systems of thought are guiding means; they are not absolute truth

—Thich Nhat Hahn

INTRODUCTION

Imagine that you are sitting with a friend in the cafeteria at the college and your friend identifies a classmate in the Tim Hortons' line-up as the "Canadian guy." You may have learned that First Nations people were the original inhabitants of Canada, or that Black Loyalists have been in Canada since the 1800s. But in spite of this knowledge, you would probably look out for a white Canadian. Where does your assumption come from? What factors have shaped your ideas, attitudes, and values? These are precisely the kinds of questions we will address in this chapter.

As we go about our daily activities we are engaged in a web of relationships and activities that connect us to the larger world. As we navigate through various interactions, make sense of the world and our place in it, we rely on ideas and values to form opinions, make assumptions, and arrive at conclusions. However, many of us don't have the faintest idea where these ideas originate from or how they influence our thinking. Most of us assume that our unexamined points of view are accurate and truthful—that they are just common sense. Perspectives that differ from our own are often dismissed, discredited, or simply misinterpreted (Paul & Elder, 2006). It is not uncommon to mistake opinions with factual analysis, ignore facts that contradict our perspective, and emphasize information favourable to our point of view. We rarely analyze and assess our perspectives in relation to alternative points of view.

This rigidity in our outlook is most apparent when discussing controversial issues where opinions tend to be polarized. Consider the exclusion of women as Catholic priests, as Muslim imams, or as rabbis in Orthodox Judaism. Individuals have different ideas about what is right and wrong with this practice. How did we formulate our particular opinions or conclusions on this issue? Can you articulate alternative views on this issue? Why is your particular view more credible than another? What premise are we using to debate the exclusion of women as a social group? Is the exclusion of women debatable in the context of equity and human rights? All that we know with some certainty is that opinions on this issue are based on religious, cultural, and political ideas humans have constructed throughout history.

In this chapter, we will think about how our ideas, assumptions, values, and perspectives are formed. We will look in detail at two case studies: youth unemployment and poverty. How do we see these issues? Do we see them as problems? If so, are they individual and personal problems? Or are they social issues that stem from and impact society as a whole? This chapter is about asking such questions—put simply, it's about engaging in a process called social analysis.

What Is Social Analysis?

Social analysis is a critical methodology that helps us to question and analyze assumptions about society. It asks us to examine the ideas that inform our perspectives and practices. It is also concerned with how these ideas are implicated in social, political, and economic factors in our society and in the larger global village. This involves observing our thinking processes, analyzing the various worldviews and ideas that inform our opinions, making connections about how these ideas shape our actions, and examining how our actions are interconnected with others at a local, national, and global level.

Social analysis could also be understood as the practice of mindfulness; it is about considering the deeper implications of the impact that ideas, actions, and behaviours can have on others, other life forms, and the environment. In this respect, social analysis enhances our ability to be self-reflective and thus mindful about the choices we make in our everyday interactions as family members, community members, and global citizens.

Social analysis is linked to social change. It requires challenging discriminatory ideas and participating in the transformation of unjust social and institutional structures. An important question for those of us who aspire to be globally conscious is: Do we unknowingly engage in the perpetuation of ideas and practices

bias

The prejudgment of others in the absence of information about them as individuals.

stereotypes

Beliefs held by individuals about the presumed physical and psychological characteristics of members of a social category.

privilege

The advantages that are awarded to those with social identities that have benefits which minoritized social identities do not.

power

The ability to construct representation of ideas or groups through the organization of meaning (e.g., whether one describes a particular armed person as a terrorist or a freedom fighter).

discourse

The boundaries within which a topic is understood and talked about.

ideology

A systematic set of beliefs, perceptions, and assumptions that provide members of a group with an understanding and an explanation of their world.

commodification

The process of reducing a person, idea, service, or relationship, not usually considered goods into an object of economic value that can be bought or sold in the marketplace.

that are harmful (i.e., maintain and create conditions of inequality, injustice, or environment destruction) or do our everyday practices work towards the promotion of fairness, inclusion, equity, human rights, justice, and ecological integrity for future generations?

Why Social Analysis?

Sociologists refer to issues as "social" because they capture the attention of a particular society at a particular time and are concerned with social interactions, organizations, and institutional arrangements that constitute our social world (Mills quoted in Mooney, Knox, Schacht, & Nelson, 2004). Regardless of where we stand on social, political environmental, economic, and global issues, we should be able to examine our thinking processes. What is the basis of the ideas and values that inform our perspectives? Where did they originate from? Can we articulate them? This process encourages us to be aware, accountable, and responsible for our thoughts and actions. It may uncover perceptual barriers such as unconscious **biases** and **stereotypes** that lead to faulty assumptions (Spellman, 1988). We all have a right to our perspective on social issues and the right to subscribe to particular worldviews or philosophies. However, this does not mean that discriminatory and hateful points of view are equally valid. The primary purpose of social analysis is to get us to think about our thinking and not to change current perspectives, though this may be an outcome.[1] Social analysis enables us to form our own opinions about social problems and how they ought to be addressed after synthesizing information and divergent perspectives from a range of sources.

Social action requires deep analysis and reflection; much depends on understanding concepts such as **privilege**, **power**, **discourse**, and **ideology**. Many of us want to contribute to creating justice and equity in the world. However, we need to examine our motivations and the impact of our interventions because our participation in campaigns to *help* others could in fact function to legitimize the status quo. For instance, Lyons and Hanley (2012) commenting on the popularity of global citizenship initiatives in higher education, critiques volunteer tourism initiatives as the **commodification** of social justice. And Illich (1968) drew attention to the paternalism that is inherent in any voluntary service activity. These criticisms advise us to examine our intentions when we engage in programs that promote charity and voluntary activities that do little to change systems of domination

---

1. Individuals respond in many ways when presented with information that contradicts their thinking. They may change their assumptions, they may hold on to their assumptions or they may experience cognitive dissonance and perhaps change at a later date (Moore, 1984). Cognitive dissonance is the discrepancy between what we currently believe to be true and other contradictory information. Common in social justice education as we resist things that challenge our views of self, other, and how the world operates.

and exploitation because they often mask exploitative **neocolonial** relationships between *first world* and *third world* countries and perpetuate cultural stereotypes and **dependence narratives**. Notwithstanding the positive human connections, the assistance that is received, and the self-discovery of Western participants, these exchanges rarely lead to sustained or substantial social and structural change.

Moore (1988) suggests that our social analysis and social justice efforts need to take into account structural factors or our efforts do little to change root causes of social problems. For example, in North America many die from overstress and overconsumption, whereas in other parts of the world 300,000 children die of hunger related diseases every day (Williams, 2004). Scientists tell us each year between 10,000 and 100,000 species become extinct from industrial pollution and habitat destruction and the world is on the verge of its sixth mass extinction (World Wildlife Fund, 2014). Currently we consume 30 percent more than what the biosphere can regenerate. If everyone on the planet were living the North American consumer lifestyle we would need six planets. Whereas, if everyone were living the lifestyle typical of those in Burkina Faso, West Africa, one-tenth of the planet would be sufficient (Cobb & Diaz, 2010). Numerous scholars and writers have argued that most of us are unaware of the cause of global inequality and the origins of dominant economic systems that promote harmful environmental practices (Galeano, 2000; Klein, 2007). Bigelow and Patterson (2002) point out that schools are one of the institutional structures in society that have done little to provide critical information about how certain nations became wealthy and powerful and others poor; how certain groups became privileged and others marginalized. Consequently, we view these conditions as normal, as the way it has always been, or perhaps of the individual's own making.

neocolonial
Structural arrangements where former colonial powers and current dominant powers exercise political and economic control over territories seized during colonialism for corporate exploitation of resources and labour.

dependency narratives
Stories that distort the structural basis of inequality and positions the donor as superior and benevolent.

SOCIAL PROBLEMS VS. INDIVIDUAL PROBLEMS

An individual problem is usually a matter of individual misfortune such as the loss of a job because of a breach of workplace policy. The cause and solution to this problem lies within the individual. Generally, an individual problem is experienced by few individuals. A social problem is widespread and the causes are systemic; that is, they arise from the policies and practices of societal structures. The solution to such problems is beyond the reach of the individual; it lies within institutional structures such as governments, the public sector, and corporations. For example, the working poor can be regarded as a social issue. Here, individuals are gainfully employed, however, their families are unable to make ends meet. The individuals can look for higher paying jobs. If none are available, the solution lies within government policy. Poverty reduction groups suggest that mandated minimum wage levels are insufficient to meet the basic necessities of life (Make Poverty History, 2014). Government policies such as the Foreign Worker Program are abused by corporations and they help to keep wages low. Labour policies that rely on part-time and contract workers rather than full-time jobs also help to create a class of the working poor (Ivanova, 2014).

Social problems are specific to particular societies at a particular time. What makes them social is that they have captured public attention within that society. For instance, today many regard homophobia and sexism as social problems because discriminatory ideas and stereotypes result in patterns of inequality such as unequal pay, restricted work, or higher levels of violence against members of these social groups. Nonetheless, some individuals maintain that we have dealt with discrimination and that homophobic and sexist acts are random instances of individual bad behaviour. Why is it that an issue is acknowledged as a social problem by most in society and for some it is not seen as such? How could certain issues lead to such contradictory assumptions and perspectives? A process of social construction determines which issues are seen as social problems and how they are defined and spoken about as social problems. Commonsense views or statements of truth about problems are linked to political positions or ideologies and have implications for what type of solutions or action should be undertaken to address particular problems. This is why people make such contradictory claims about social problems. Therefore, a starting point of social analysis requires that we distance ourselves from what we know about society.[2]

Our Understanding of Problems Are Socially Constructed

Social analysis requires us to be aware of our worldviews as we consider evidence and facts to establish whether problems are systemic or individual. Our values and assumptions influence how we frame issues and how we perceive the social world. To achieve some measure of objectivity, we need to examine our perspectives to understand the influences that shape our definitions of problems and the strategies we propose to deal with these problems.

---

2. For instance, many of us view the differences between girls and boys as innate, not as socially conditioned or socially constructed by ideas and practices that circulate in society. From the time boys and girls are born, we view them differently and treat them differently. Consequently, we can tell them apart by what they wear. We treat boys and girls differently based on socially defined expectations of appropriate gender roles and behaviours. We expect them to behave and act in particular ways. Generally, we expect girls to be more passive than boys. Girls get dishes and dolls, and boys weapons and trucks. Boys and girls internalize these social messages from family, media, and the education system. When they act accordingly we reward these behaviours and see them as normal. However, these differences have been socially and culturally produced. Not all girls and boys confirm to socially produced gender norms. Those that don't conform are stigmatized as "gay" or "butch." This is an example of the social construct of gender (Lorber, 1994). What it means to be a boy or girl in society has been constructed by a system of ideas and practices that individuals are expected to conform to; these are called norms. Individuals that don't conform to socially constructed norms can be stigmatized because they challenge dominant assumptions.

Knowledge is not neutral; it is socially constructed within historical, cultural, social, and political contexts. It is also mediated by relations of power; it can be shaped by the powerful as well as those who are marginalized and exploited. Knowledge, what we know to be truth or common sense, is highly variable. Therefore certain bodies of knowledge are constantly being shaped and reshaped. Historically, scientific certainties and sociological and psychological theories have been challenged and reevaluated. Feminist and antiracist historians, for example, have pointed out that recent history is told from the perspective of powerful white men excluding other perspectives. We know a great deal about Columbus but very little about the people whose land he took over and were massacred. We celebrate Columbus as having discovered North America when there were hundreds of indigenous nations already occupying this land. Elementary students learn about the heroic acts of European settlers but very little about the experiences of First Nations who experienced cultural genocide and deaths in residential schools.

Even when presented with information that contradicts our position, some of us hold on to our perspective. If we take homelessness as an example, some will see the individual (the homeless person) as the problem even when presented with overwhelming information that this is a systemic problem, in other words, the result of corporate and government policy. Cutbacks by governments over the last two decades to social and mental health services, the lack of affordable housing developments, the corporate gentrification of slum areas, and the lack of government policy on affordable housing are tied to the rise in homelessness. Because of cutbacks to social welfare, even single parents with children are now experiencing homelessness and extreme poverty. These political decisions are influenced by ideas about the poor (Make Poverty History, 2014). Discriminatory ideas and stereotypes that stigmatize certain social identity groups may cause us to overlook systemic patterns. This encourages us to blame individuals for social problems and distracts us from the source of these problems. Consequently, many perceive homeless individuals as a blight on the urban landscape. For these individuals the homeless are threatening or bothersome; they clutter the streets, may be potentially violent, and may carry disease. They may even want the authorities to remove the homeless from sight or incarcerate them and restore "law and order."

Historical, Economic, Political, and Cultural Variables Shape Our Understanding of Problems

Social analysis requires that we understand how social problems are shaped by historical, economic, political, and cultural variables. Let's take unemployment as an example. Underemployment was seen as an individual problem in the 18th century and as a social problem in the 1940s. Today, it tends to be understood again as an individual problem. Political, economic, and historical circumstances have influenced the dominant or commonsense view of this problem.

The 18th and 19th Centuries

In the 18th and 19th centuries in Europe, the poor were stigmatized and poverty was regarded as an individual problem. During the transition from feudalism to industrialization, which also gave rise to the economic system of **capitalism**,[3] many peasants were displaced by the enclosures of land previously held in common[4] and sought work in urban areas. This economic activity involved the use of unskilled, underpaid workers and horrific working conditions. Many endured long periods of underemployment. This was also the case in North America where Europeans were forcibly displacing the First Nations and colonizing Canada. Control of these territories required workers and many poor Europeans immigrated to work in the burgeoning industrial economy. Since working conditions were similar to those in Europe, labour strife resulted in strikes for better working conditions and better pay. Workers did not have health care; illness and the lack of funds for medical care meant certain death. There was no vacation time off or limits on the work day and work week. A worker could work 14 hours a day, 7 days a week. These working conditions were considered the norm (Camfield, 2011).

The 1940s

Heightened tensions created by economic and political developments such as the Great Depression, union activism, and the perceived threat of communism[5] led to a new way of understanding unemployment. What was previously seen as an individual problem in the 1940s was now understood as systemic, created by economic structures and policies beyond the individual's control. Union activism and social criticisms of the capitalist economic system led to a new understanding about the unemployed. Under capitalism, about five to 10 percent of working age people would never find employment.[6] Furthermore, workers were underpaid because industrial capitalism is designed to maximize profits through the exploitation of workers by devaluing their labour. Rather than ferment unrest in the turbulent

capitalism
A global economic system characterized by the private ownership of the means of production/private property. The capitalists' main aim is to produce goods to sell at a profit by keeping the cost of labour and resources low.

---

3. An economic system characterized by the private ownership of the means of production. These owners are called capitalists and their main aim is to produce goods to sell at a profit. During the early phase of industrialization, the major capitalist was the factory owner. More recently, merchant capitalists deal specifically in the buying and selling of wholesale commodities at the international level. Under this system there is never full employment for the population to account for fluctuations in the economic system. To mitigate against these conditions, the social welfare system was established.

4. In the 18th and 19th centuries, the English government and aristocracy created privately enclosed plots of land formerly owned in common by all members of a village or at least available to the public for grazing animals and growing food. The aristocrats who now owned the legally confiscated land were able to profit from commercial agriculture.

5. Communism is an economic system based on the ownership of all property by the community as a whole.

6. Full employment can only exist under some state-controlled socialist economies. The capitalist economic system is unstable with fluctuations in employment levels; the unemployed function as a reserve army of labour to account for this uncertainty.

political climate where socialism[7] might become an attractive alternative, an agreement was struck among labour, the capitalist class, and government. Western capitalist countries that adopted these socialist-inspired humanitarian programs to address structural inequality were referred to as **social democratic** states. Social assistance provided relief for those who could not find employment, unemployment insurance for the temporarily unemployed, collective agreements for workers to bargain for better wages, health and safety legislation to protect workers, limits on the work day and work week, among other measures to protect workers from exploitative practices.

> **social democratic**
> A political perspective that includes aspects of liberalism and socialism. Adherents believe government should promote the collective good and play a productive role in the economy to bring about greater equality and distribution of resources and increase democracy.

The 1970s to the Present Day

From the 1940s to the 1970s, the height of the industrial revolution, this accommodation among capitalist elites, unions, and government resulted in stable well-paying jobs and social policies in times of need. However, this changed when new ways of framing the problem emerged in the late 1970s that attacked the concepts of social assistance and human rights.[8] Since these developments, the dominant perception of unemployment and poverty as individual problems has reemerged. The commonsense assumption is, if you really want to get a job you can find one. Therefore it is the individual's fault if they experience unemployment and poverty. Stereotypes were deployed by right-wing politicians characterizing the poor as unmotivated, lazy, and living the "good life" as manufactured in the caricature of the "welfare queen."[9]

Resistance Can Change the Social Construction of Problems

To understand the ways in which people can organize to change the definition of a social problem, we might turn briefly to the case of South Africa. For decades after the United Nations *Universal Declaration of Human Rights*, the Apartheid (white supremacist) government of South Africa continued to function within the global community with support from powerful heads of state and governments. For example, U.S. President Ronald Reagan, British Prime Minister Margaret Thatcher, and Stephen Harper (not yet Canada's prime minister) were among those who vilified Nelson Mandela as a terrorist when he was leading the struggle against the white supremacist Apartheid regime in South Africa. These leaders of some of the

---

7. Socialism is an economic system where ownership and operation of the means of production and distribution as well as work is shared by the community rather than by private individuals.
8. Such as neoconservatism and neoliberalism, discussed later on (see Figure 3.1 on page 47).
9. A term, popularized by Ronald Reagan, which is regarded as a racist and classist stereotype to describe racialized women who live "lavish" lifestyles off the welfare system by having many children and creating many identities to increase their monthly payments.

most powerful nations in the world did not see legalized and institutionalized state racism as a social problem (Saul, 1986).

The views of these prominent Western politicians reflected tensions over the struggles to advance equity and human rights within their respective countries. Although many racialized and white individuals in the West did not support this ideology, they unwittingly participated in economic structures that supported this regime. Canadians dealt with commercial banks and companies that had investments in Apartheid South Africa and therefore helped to prop up the regime. The shift from Apartheid being ignored to becoming an international social problem in urgent need of change, came from below. Through the efforts of university students and professors working as allies with anti-Apartheid activists, there was increased awareness and condemnation in the West. The divestment campaign began to shut down the South African economy and Apartheid was now seen as a social problem that was morally indefensible.

In the political context that has evolved since the anti-Apartheid movement, Mandela is now viewed as a freedom fighter and even embraced by government leaders who suppress human rights and deny rights to **minoritized groups** within their own countries.[10] In addition to the personal virtues of Mandela himself, there are a number of complex variables that explain why the view of Apartheid and Mandela has changed over time.[11]

As these examples illustrate, the perception of problems as being individual or social can change over time because of activism or social, historical, and economic variables. What these examples also illustrate is that the definitions of problems are always contested especially where conditions of inequality and injustice exist. Those who are oppressed and their allies can and have changed the way issues are defined. However, those in power have more means to deflect or misinterpret social problems when it threatens their interests. In the next section we will examine some of the reasons why people have vastly different views on issues.

minoritized groups
This term results from discriminatory discourses about oppressed social groups that become commonsense assumptions that circulate within the larger population and can also be internalized by members of the targeted group.

10. Canada gives Aboriginal children half of the funding of other Canadian children. This differential treatment is similar to Apartheid policies (http://www.cbc.ca/thesundayedition/shows/2013/01/13/former-south-african-ambassador-on-aboriginals-in-canada/) (Boyden, 2013). A little known fact is that the South African government came to Canada in the 1900s to study the reservation system from which apartheid was developed. Canada was one of the first western countries to challenge Apartheid after citizens' groups and university student groups waged decades-long campaigns against the Apartheid regime (http://www.thestar.com/opinion/2013/12/11/canadians_did_more_than_anyone_for_south_africa_siddiqui.htm) (Siddiqui, 2013). Stephen Harper (before he was Canada's prime minister) did not support Mandela when he was in prison. See http://rabble.ca/columnists/2013/12/stephen-harper-nelson-mandela-and-whitewashing-past

11. Many corporations were concerned that Mandela would nationalize the economy since he embraced neoliberalism. Though South Africans have political freedom, many still do not have freedom from the extreme poverty that existed under Apartheid.

DEVELOPING CRITICAL METHODS OF ANALYSIS

> In a world where justice is mal-distributed there is no such thing as a neutral or representative recapitulation of the facts.
>
> —Howard Zinn

Social analysis requires that we examine how social problems come to be characterized as personal or social by identifying various perspectives and examining commonsense assumptions associated with these points of view. Several questions arise when doing this kind of analysis:

- How do issues come to be seen as social or individual?
- What are different ways of framing an issue?
- Who stands to benefit from a particular framing of the issue?
- Who is negatively affected?
- What ideas, values, or interests are being promoted by various ways of framing a problem?
- How is language used to discuss and characterize the issue?
- Is significant information omitted?
- What additional information could provide a more complete understanding of the issue or problem?
- Who or what are the sources generating different perspectives?
- Are they powerful elites or disenfranchised and minoritized groups and their allies?
- Before deciding that an issue is personal (an individual flaw) have we considered social (systemic) variables?
- Are there societal impacts associated with framing the issue as an individual problem? (Swift, Davies, Clark, & Czerny, 2003).

These basic critical thinking questions help to make the familiar unfamiliar; that is, they help us to lay bare that which we often take for granted. They encourage us to think about the operation of power by systematically probing deeper into the generation of perspectives and frames of reference used to define and discuss issues (discourses), the interests behind these perspectives (**hegemony**), and the systems of ideas (ideologies) we adopt as commonsense explanations to make sense of the world.[12] As we shall see, these questions and concepts are the foundational building blocks of social analysis.

hegemony
Dominance is not achieved through direct authoritarian rule, but through a process of building consent through social practices where the ruling classes present their interests as the general interests of the society as a whole.

---

12. We will explore the concept of ideology further on. Ideologies are essentially competing ideas based on systems of thought about how society ought to function. Propaganda is a tool used by proponents of various ideologies to distort and/or omit facts to validate a particular perspective.

Three Approaches to Framing Social Problems

This section illustrates three ways in which social problems might be framed or socially constructed. First, they could be read as individual problems. Second, they might be understood as collective problems that are caused by the failure of public institutions. Third, they can be understood as systemic problems, that is, as problems linked to the very historical, political, social, and economic systems in which they operate. To explore these three frameworks, we will look at youth unemployment.

Youth unemployment has been increasing at a rapid rate in North America and Europe.[13] Most individuals have an opinion on why youth are unemployed, but if one were to conduct a survey on this issue there would be no consensus. However, people will speak about this issue and its causes with certainty.

Youth Unemployment As An Individual Problem

1. "Today's youth are entitled and lazy, they expect too much." "We must take for granted that good jobs are a thing of the past, the new reality is that work is unstable, low paying, without benefits or pensions." "There are jobs out there; if you try really hard you will find work."

Youth Unemployment, A Problem Caused by the Failure of Public Institutions

2. "Students are being shortchanged by the educational system. Schools don't adequately prepare students with appropriate skills and knowledge for the job market."

Youth Unemployment As A Social (Systemic) Issue

3. "Young people are underemployed in many sectors of the economy because government policies favour profits for the corporate sector at the expense of the general public. Contributing factors include government cutbacks and outsourcing along with corporate downsizing, deskilling and union bashing. Hence, low paying, unstable and temporary jobs where many people and particularly youth are vulnerable to becoming the working poor."

---

13. Ontario's monthly youth employment rate—a measure that determines how many young people actually have jobs—fluctuated between 50 to 52 percent cent, meaning half of all Ontario youth don't have jobs. Read more: http://www.ctvnews.ca/canada/ontario-youth-unemployment-among-the-worst-in-canada-report-1.1473423#ixzz2moi8OEDx
See also: http://www.telegraph.co.uk/finance/jobs/youth-unemployment-competition/9850404/The-winning-essay-How-to-solve-youth-unemployment.html
http://rabble.ca/news/2013/09/young-and-jobless-new-ccpa-report-examines-youth-unemployment
http://bilbo.economicoutlook.net/blog/?p=24157
http://rabble.ca/news/2013/10/jobless-not-broken-youth-workers-gather-to-talk-unemployment

To see unemployment as an individual problem is to assume that, "if you try hard enough you can find a job." Therefore if you are unemployed, the assumption is that it's your own fault. The few that subscribe to the latter views are those likely to have experienced looking for and not finding work, and those who know highly skilled and motivated individuals who at some point in their careers are simply unable to find work. The self-evident conclusion in the first assumption is that there will always be people who won't help themselves because in this society anyone can get a job if they try hard enough. Indeed, there are lazy and unmotivated youth today as there have been in every generation.

While the first perspective blames youth unemployment entirely on the individual, the second perspective suggests that public institutions, in this case the educational systems, are partly to blame for the problem. Both of these perspectives are problematic. The first argument—youth unemployment is an individual problem—contradicts the facts. The youth unemployment rate is usually 3.5 times that of the adult population. In 1998, the unemployment rate was 8.3 percent for Canada's adult population; it was 28.7 percent for youth and youth unemployment has steadily increased (Klein, 1999). Youth are also twice as likely as adults to be laid off and experience unstable employment. In 2012, the majority of youth who were employed one month and unemployed the next were laid off.[14]

To perceive youth unemployment as an individual problem is to overlook the ways in which the current economic system feeds into the problem. As stated earlier, in North America we live in a capitalist economy, this economic system does not have full employment. Technically, full employment is only found in socialist economies where the state has central control over economic decision making (Wallersein, 1989). In capitalist economies, there will always be fluctuations in the labour market based on factors such as production, consumption, imports, exports, the price of the dollar, competition, inflation, speculation, among other variables. This uncertainty is accommodated for with a "reserve army of labour," a permanent pool of unemployed people who will not be able to find jobs at any given time (Braverman, 1998). The economy is considered healthy if the unemployment figure hovers around 5 percent to 10 percent. Currently, in some European countries the unemployment rates are as high as 26 percent.[15] Thus, a pool of individuals is always available to fill high demand jobs during times of economic fluctuations. This ensures that business needs are not disrupted. Historically, corporations and businesses use this economic structure as a power leverage to keep the price of labour low; this uncertainty in the labour market historically kept the power of labour unions in check. If there was full or close to full employment, workers could bargain or organize for better wages. More recently, corporations hire temporary

14. http://www.statcan.gc.ca/pub/11-626-x/11-626-x2013024-eng.pdf
15. http://epp.eurostat.ec.europa.eu/statistics_explained/index.php/Unemployment_statistics

foreign workers, through a federal government program that displaces youth workers, drives wages down, and creates exploitative conditions for foreign workers (Block, 2014; Canadian Press, 2014). The evidence suggests a structural rather than an individual problem in this instance.

We can conclude that to call youth unemployment an individual issue is an example of a commonsense assumption that is presented as a fact which contributes to the stereotyping and stigmatizing of unemployed youth. Such assumptions serve to deflect attention away from a systemic problem. As well, the lack of government policy initiative in this area can obscure the flaws with the economic system (Olive, 2014).

The second assumption—that youth unemployment is caused by public institutions—takes into account the fact that educational institutions often fail to prepare students for the job market. Indeed, educational institutions could shoulder some of the blame. Some diplomas and program offerings may not always lead to meaningful jobs and some institutions do not provide the relevant skills necessary in a competitive job market. However, the problem here is that these concerns are related more to employment preparation than to the root causes of unemployment. While university and college education is extremely costly and may not be accessible to all, the problem is not strictly the institution itself but the chronic government underfunding and deregulation of fees.

The second assumption, like the first one, also fails to consider the connection between youth unemployment and the current economic system. This economic system is not creating more opportunities for the middle and working classes in society. In fact, the evidence tells us it is creating conditions of increasing inequality (Stanford, 2014). Since youth are among the most susceptible groups in the workforce, inequality and lack of opportunities have a major impact on their ability to find jobs and become self-sustaining. Blaming educational institutions for youth unemployment deflects attention from the structural causes of this problem. Public institutions were created to provide services to the public that ensure access and equity to health care, education, and other services. Smith (2004) points out that since the 1970s there has been a concerted attack on these institutions as burdensome expenditures that create public debt by neoconservative interests. These interest groups and politicians have constructed a discourse that frames public institutions as inefficient, even unnecessary, and promotes the idea that public institutions should operate as though they are a business selling a product to consumers. Critics of this perspective see it as a larger trend towards prioritizing the needs of the wealthy and draw attention to trends such as the privitization[16] of

---

16. Privatization is the transferring of public services such as health care, utilities, education, and prisons to private for profit corporations in the belief that it will lead to more efficiencies. Critics claim that the evidence does not back up this approach and leads to less equitable access.

public services. In this ideology, the role of public institutions is not to create conditions of social justice or protect the environment, but to liberalize markets for wealth accumulation. The health of financial markets that primarily benefit the rich takes precedence over other indicators of a healthy society such as good stable employment, access to affordable education, and access to social services for those in need (Truscello, 2011). Under this shift, education is less about building a more equitable, just, and sustainable society and more about serving the priorities and interests of those that control economic and financial markets. Ritzer (1995) suggests that this corporate rationalization of all human activity prepares us for a shift in the structural organization of society that is increasingly concerned with efficiency and formalized control of workers and public institutions.

The third perspective argues that we need to examine the root causes of such problems by focusing on structural and systemic factors and locating these historically. Adherents of this perspective maintain that we cannot overlook corporate influence on government policy that has resulted in growing inequality, less stable job prospects, and related problems such as environmental destruction.

Historical employment data indicates that employment levels were relatively stable in North America from the 1940s to the 1970s during the peak of the industrial revolution. Labour unions and political developments influenced government to adopt socialist inspired collectivist policies such as fair wages, safer working conditions, and the provision of social services for those inevitably disadvantaged by the capitalist economic system. This combination of capitalist and socialist policies is referred to as social democracy (Macridis & Hulliung, 1996). At the time, this accommodation balanced corporate needs for maximizing profit and citizens' rights to earn a fair wage for their labour. However, the 1970s saw a reemergence of a pro-corporate agenda that sought to gain more control of the economy and lessen the size and role of government. These developments converged with the rise of religious fundamentalist groups that attempted to influence government policy with the promotion of social and moral conservative values. These ideas, which we will examine in more detail in the next chapter, were known as neoliberalism and neoconservatism[17] (Smith, 2004).

Economic data indicates that inequality continues to increase; the middle class is eroding into the ranks of the poor and underemployed (Yalnizyan, 2010). Even as corporations rake in massive profits they continue to downsize employees, ship jobs overseas to sweat shops, or replace them with low pay contingency workers. Over 3 million Canadians live below the poverty line. These include not only the underemployed or unemployable but also the working poor. Employment poli-

---

17. Neoconservative and neoliberal economic policies are similar; both favour corporate interests. Their primary difference is in their approach to social problems. See Figure 3.1 on page 47 and the next section on ideology.

cies have led to the de-skilling of jobs, the reducing of full-time jobs, and the disbanding of unions. Policies to undermine unions and job protection have been very successful in Canada—only 11 percent of the workforce is now unionized, most of these in the public sector (Mills & Simmons, 1999).

The concentration of wealth continues to increase. The wealthy continue to increase their share of the economy while the poor increase their debt load. Not only that, employment demographics are also highly racialized and gendered. There are four to five times more racialized families living below the poverty line than white families and 60 percent of minimum wage jobs are held by women (Nestle & Kanee, 2008). McQuaig and Brooks (2012) suggest there is every indication that this trend towards a vast disparity in wealth between the haves and have nots will continue. Not only are there fewer jobs, youth are working in jobs that underutilize their skill levels and are forced to settle for part-time, temporary, low-wage jobs (Goar, 2014).

Apple (1993) points out that pro-business interests argue for cutbacks to government, public institutions, and social programs that benefit citizens, but they welcome the government's healthy corporate subsidies, grants, and policies that advance their interests. North American governments spend up to four times more on *corporate welfare*[18] than what they spend on social welfare.[19] In Canada, oil sands-based corporations receive approximately $3 billion in corporate welfare from the government (Nelson, 2012). The corporations and governments claim that corporate tax cuts, grants, and special loans are necessary for correcting market failures. The practice of giving taxpayers' money to businesses defies the rules of capitalism. Under a free enterprise system, businesses succeed or fail based on the open marketplace.[20]

Over the last 40 years, governments have steadily cut corporate income taxes. In Canada the corporate tax rate in 2000 was 28 percent and then in 2012 it dropped to 15 percent.[21] Despite $14.2 billion in worldwide profits, General Electric and

---

18. This is a term used to describe government financial aid, such as a subsidies, tax breaks, and grants to corporations and other private sector businesses especially when viewed as wasteful or unjust.

19. Huff's (1993) research on corporate welfare revealed that in 1990 the U.S. federal government spent $170 billion on corporate welfare. In that same year $11 billion was allocated to Aid for Dependent Children, Medicaid, the most expensive welfare program received $30 billion, $4.7 billion was spent on international aid, pollution control programs received $4.8 billion, both secondary and elementary education were allocated $8.4 billion, and the federal program for the disabled received $13 billion.

20. Adam Smith, the father of capitalism, in his treatise *The Wealth of Nations* stated that "invisible hand" or self-regulating behaviour of the marketplace is what should allow individuals to make a profit, and maximize it without the need for government intervention. The "invisible hand" of the marketplace sorts out the profitable from the unprofitable, under no circumstances should there be government intervention.

21. http://www.policyalternatives.ca/sites/default/files/uploads/publications/National%20Office/2011/04/Corporate%20Income%20Taxes,%20Profit,%20and%20Employment.pdf

numerous corporations did not owe taxes in 2010.[22] In Canada, corporations made 52 percent more profit in 2009 than in 2000, yet they are paying almost 20 percent less in federal/provincial income tax (MacDonald, 2014).

When the pro-business lobby is criticized for this apparent double standard, they claim that subsidies for public institutions and social programs such as Medicare and social welfare are examples of socialism and should be cut back or eliminated. Since the 1980s, welfare payments have been cut back to such an extent that most recipients have to rely on food banks to make ends meet[23] (Goodman, 2014). In addition, funding continues to be cut back for public institutions such as health care, social services, transportation, and education. University and college tuition has increased; tuition for degrees that lead to well-paying jobs has been deregulated. Law and medical school tuition is so high that it is beyond the reach of most middle and working class students. In addition, many universities and colleges have been forced to get private sector funding for certain programs, which concern many in academia because of the fears of corporate control of education and research. These developments have led critics such as Parenti (2014) to suggest that this amounts to "Socialism for the rich and capitalism for the poor."

Proponents of the third argument suggest that the collusion between governments and corporations that result in growing inequality and unstable employment is a direct consequence of **crony capitalism**. Much worse, democratic gains have been eroding because corporations use their wealth and paid lobbyists to influence the democratic decision making process. Klein (2007) points out that since the 1970s, lobbying to dismantle policies that protect workers and the environment by framing them as unnecessary red tape that reduces profits and slows down the economy has been very successful. Over the last two decades, governments have deregulated health and safety standards to protect workers, weakened employment and pension policies, and deregulated environmental protection legislation in favour of industry self-regulation. This has resulted in a number of public disasters such as the Walkerton crisis, the Maple Leaf Foods Listeriosis outbreak, and the Lac Megantic rail disaster (Hennessy, 2014).

Many now see corporate interests as benefiting the public interest, having bought into the pro-corporate framing of what is best for the private sector is good for the economy and everyone else. Few question the massive cuts to public institutions, many now see them as bloated and inefficient with overprotected and overpaid unionized public servants. The values of competitive individualism promoted by dominant perspectives have been adopted by many in a *race to the bottom*—if I

crony capitalism
An economy characterized by favouritism in the distribution of legal permits, government grants, special tax breaks, lax environmental policy, and other state intervention in favour of business interests.

---

22. http://www.businesspundit.com/25-corporations-that-pay-less-taxes-than-you-do/
23. http://www.truthdig.com/report/item/congress_to_the_unemployed_eat_confetti_20131231

can't have a good job, why should you? Human rights activists, environmentalists, and even scientists are labelled radical when they challenge the negative impacts of government and corporate collusion.[24] Instead of making governments more accountable to citizens, we attack unions which, for many decades, were able to temper the exploitative practices commonplace under the economic system of capitalism.

The process of critical inquiry presented above used the case study of youth unemployment and unemployment in general to pose the question, is this a social or an individual problem? As we demonstrate there are a number of ways in which social problems are framed or socially constructed. Those who have the power to access mechanisms of communications such as educational systems or mainstream media and entertainment industries are in a position to influence public opinion. They can frame problems to conform to a particular set of ideas (ideologies). Usually, though not in all cases, this becomes the dominant perspective (McChesney & Nichols, 2002). Even though particular ways of framing issues may be based on erroneous assumptions, inaccurate and incomplete information, they have the potential to become taken for granted. Repeated as truth, they become dominant perspectives circulated in the everyday discourse in the corporate mainstream news media, political talk shows, and in most business courses in college or university. Chomsky and Herman (1988) suggest that this is a method of "manufacturing consent" within the population (see Chapter 5). The ideas and opinions that benefit the elites (such as values of individualism over collectivism) are uncritically consumed by most of the general public as truth, including those negatively affected by these economic and political structures. As a result it becomes the norm to blame social problems such as homelessness, underemployment, and other forms of inequality on the individual rather than the structural economic and political roots of the problem (McQuaig & Brooks, 2012).

Counter discourses[25] by student groups, unions, and anti-poverty groups are less likely to shape the understanding of structural inequality because they don't have the same access to communications mechanisms as corporations, nor are they called on to speak to the issues as experts. Consequently, they are less likely to shape what becomes the commonsense social construction of the issue. However, not everyone is a passive recipient of the dominant perspectives or ideologies. While dominant perspectives are consciously shaped to promote particular interests they are also contested. We see this with mass social moments such as unionism in the last century, the gay liberation movement in the 1960s, the anti-Apartheid movement in

---

24. See: http://sciencewriters.ca/initiatives/muzzling_canadian_federal_scientists/ and http://www
.vancouverobserver.com/sustainability/government-labels-environmentalists-terrorist-threat-new-report
25. Dominant discourses are constructed within relations of power and therefore are always subject to challenges regarding the framing and understanding of issues.

the 1980s, and more recently with movements like Occupy Wall Street, Idle No More, Cochabamba, and the Arab Spring. Historically inequality, oppression, and injustice always lead to economic, political, and social instability. Consequently, dominant discourses are always reinvented as they are discredited or exposed as discriminatory, oppressive, or unjust (Salih, 2002).

Important Concepts for Analyzing Social Problems

Discourse

The above case study illustrates three distinct ways of talking about the issue of youth unemployment. Each approach to framing the issue and drawing conclusions about the cause of the problem represents a particular discourse. Discourse is a concept that explains how knowledge is produced about various topics; it provides a language for representing the topic and talking about it. Brock (2003) describes discourse as the way in which power, language, and institutional practices combine at historically specific points to produce particular systems of knowledge that influence how we think and speak about various issues and social structures.

According to Mullaly (2010), discourses are not neutral; rather, they are delivery systems for political and social assumptions about the world that reflect the interests of powerful groups such as Euro-North Americans and wealthy males and assist in reproducing existing class, race, gender, and other inequalities. This reproduction occurs through the promotion of ideas that supports the current social and political order and by suppressing ideas that critique present arrangements or that try to transform them. The first two discourses in the previous case study represent dominant discourses about youth unemployment. Alternative perspectives such as the perspectives (counter discourses) presented in the third reading of this social problem are generally dismissed as irrelevant, inappropriate, radical, and unreasonable (Henry & Tator, 2002).

The repetition of dominant discourses about the current economic system that creates and maintains inequality explains why many of us view corporate interests as benefitting our own interests when there is evidence to suggest otherwise. Powerful interests are able to define what we know about the economic system and establish the terms regarding how this issue is understood and discussed. Alternative perspectives and evidence that challenge these dominant discourses are suppressed or discredited. Consequently, we don't recognize that discourse is a form of power that operates through ideas, language, professional practices, and institutions to create taken-for-granted *truth* or norms.

Ideology

Discourse is similar to the concept of ideology, but whereas discourse functions to define what we know about certain topics, ideologies articulate a vision of how the world or society ought to be organized and promote a values framework. Ideologies serve the interests of a particular class or social group by obscuring social, historical, and political processes that make everyday life seem natural and unchangeable. For instance, many of us see the economic system of capitalism as the only rational form of economic organization when it was created to benefit particular classes and groups through human activity and conflict over several centuries (O'Brien & Szeman, 2010). Many of our beliefs about cultural, social, economic, and political arrangements are derived from ideologies that we subscribe to. We have very firm beliefs about these arrangements; however we do not view them as historically and socially constructed by ruling elites for their benefit. We take these beliefs for granted, unaware that they are driven by systems of ideas or ideologies that inform our worldviews (Macridis & Hulliung, 1996). By identifying specific ideologies, we are able to name social and political processes that directly and indirectly hide the historical context behind ideas and structures that appear as common sense or natural. Dominant discourses and ideologies function to legitimize the present organization of society and rationalize negative social and economic impacts experienced by marginalized groups.

Hegemony

The promotion of dominant discourses, ideologies, and coercive practices by ruling elites to manufacture consent among the population is referred to as hegemony. According to Mullaly (2010) hegemony is achieved through the control of the education system, religious institutions, and the mass media. Furthermore, power is maintained internationally through economic and cultural imperialism. To maintain the legitimacy of the elite's position of social dominance, consent must be actively maintained because the social and political power that benefits the few and creates inequality and oppression is always challenged by alternative discourses and ideologies.

Ideologies are used by powerful elites to legitimize their dominance and enlist the support of the larger society. This is achieved by promoting the belief that the social hierarchy is necessary for the well-being of the society as a whole and that the current social structure is inevitable and natural. This is intended to enable subordinate social groups to feel that existing relations of power are legitimate and normal. Through these processes, dominant discourses and ideologies become invisible—assumptions about the world—and dominant elites need not use physical force to rule since they build consensus through positions of intellectual and moral leadership. Hence, a commonsense view of the world is inherited from the past and uncritically absorbed producing a moral and political passivity. Despite exploitation and oppression, the subordinate groups see the interests of the ruling class as the general interests of all groups in the society (Storey, 2009).

How Ideologies Influence Our Perspectives

> The poverty of our century is unlike that of any other. It is not, as poverty was before, the result of natural scarcity, but of a set of priorities imposed upon the rest of the world by the rich. Consequently, the modern poor are not pitied... but written off as trash
>
> —John Berger

The next case study illustrates how we all use ideology in our everyday thinking. Ideology is an integral part of our values framework. We use these systems of ideas and patterns of thinking to make sense of the world, but, they are often invisible to us.

Take a look at the causes of poverty in Figure 3.1. Highlight the ones that match your own opinion or that you have heard on television, from family and friends, and what you may have read in textbooks or newspapers.

FIGURE 3.1   What Are the Causes of Poverty in Canada?

A	B	C
Laziness—anyone can get a job	Social services are inefficient at helping the "disadvantaged"—retraining, getting out of the "culture of poverty"	Social, political, and economic structures/policy cause inequality and poverty
Abuse of the welfare system		Government cutbacks on social services
Drug addiction	People can't manage their debts	
Single mother families		More corporate welfare than social welfare; lower corporate taxes or no taxes
Lack of ambition	School is too expensive; we need a better loan system	
Dropping out of school		Attack on the idea of unions
Bad choices	Public education system is not relevant—leads to high dropout rates	Low paying insecure jobs; low minimum wage below inflation and poverty line
		Corporatism—policies that favour the rich and corporations at the expense of the poor and middle class
		Inequality that is racialized, gendered, and based on ability/disability

Columns A, B, and C represent generalized ideological views on the root causes of poverty. Column A represents a neoconservative ideological view and Column B is representative of the neoliberal ideology; these are the dominant political ideologies today. These are also referred to as right-wing or pro-corporate ideologies. Column C is associated with various progressive or transformative political and social ideologies that are labelled as left-wing ideologies. Since individuals are complex we don't fit neatly into boxes, we may subscribe to more than one ideology depending on the interplay between our histories, social identities, the particular issue at hand, and the specific context that is pertinent.[26] We will trace the history of these ideologies and their present day manifestations in the next section.

If most of the causes you agreed with are located in Column A, your views of poverty fit a pattern that characterizes poverty as an individual problem; this aligns your thinking with a neoconservative ideological outlook. Poverty is caused by the individual's attitudes and values that lead them to make poor choices. Poverty then is an individual failure not a societal one. It has nothing to do with how the society is organized. Poverty has nothing to do with social structures that make some wealthy and others disadvantaged. Since poverty is not a social issue, it does not require systemic interventions such as social welfare; the poor should rely on charity. It is not the responsibility for the state to provide for those who are poor. This is one of the dominant discourses of poverty one hears in North America. Since North America is the "land of opportunity," if you are poor in this society, it is because of a personal flaw. This framing of the issue situates the poor as a problem to be dealt with. Consequently, the focus is on how best to manage and police poor people: How can we stop welfare cheats and abusers of the system; how can we manage the poor so they don't disrupt the social order. The use of language such as "welfare queen" dehumanizes and stereotypes the poor, which makes it easier to cut or eliminate social services that provide for the marginalized.

Column B responses acknowledge that the structure of the society is a source of poverty, but the personal deficiency view is not dropped entirely. This social and individual problem is framed as a naturally occurring aspect of the social and economic system. This ideology is referred to as neoliberalism. It supports existing arrangements of power, so the issue is how best to contain and manage the poor. It supports institutional programs to help those who become "disadvantaged" by the system. The focus is on how best to develop social work and child and youth worker programs to help poor people. Attention is then deflected from the under-

---

26. It should be noted that there are many more layers to this analysis of poverty and inequality. A more in-depth analysis would uncover that there are patterns of poverty along lines of race, gender, and disability, and other social identity groups. We would then examine the various explanations and ways of framing this social problem. We need not stop here; we could also examine patterns of poverty between the global north and south, and look at the various explanations and rationalizations.

lying structures that are responsible for creating inequality in the first place. The problem becomes the poor and the institutional programs developed to *help* them. This framing sees poverty as normal in society. It then places the onus on the poor to escape poverty by taking advantage of services that better equip them to function in a competitive society where the fittest of the fit survive. Another source of the problem is the function of public institutions. Questions may be raised about their relevance: Are they efficient and effective, or should they be cut back, privatized, or eliminated?

Those who identify primarily with Column C focus on systemic failures of the existing economic system as the primary cause of poverty. This approach is associated with transformative ideologies that focus on systemic explanations as the root cause of poverty. Advocates of this approach argue that the system cannot be reformed, what is needed is fundamental change in the organization of society because it is exploitative and unfair. This view suggests that the system is structured to advantage the wealthy and powerful.

What will become apparent from the case studies presented in the next chapter and our discussion of the major ideologies is that dominant ideologies and dominant discourses are interrelated. Discursive patterns of framing issues to produce *truth* are closely associated with particular ideologies.

CRITICAL THINKING QUESTIONS

1. How are commonsense assumptions formulated?
2. What makes an ideology or opinion appear more credible than others?
3. How do we unwittingly engage in the perpetuation of harmful ideas and practices?
4. Why is it that an issue is acknowledged as a social problem by some in society and for some it is seen as an individual problem?
5. What processes lead to such contradictory assumptions and perspectives?
6. What are different ways of framing an issue such as poverty?
7. Who stands to benefit and who is negatively affected by particular ways of framing this issue?
8. What ideas, values, or interests are being promoted by various ways of framing the issue of poverty?
9. How do various perspectives use language to discuss and characterize this issue?
10. What information has been omitted from particular ways of framing the issue?
11. What additional information could provide a more complete understanding of this issue?
12. Who or what are the sources generating different perspectives?

REFERENCES

Apple, M. (1993). Constructing the "other": Rightist reconstructions of common sense. In C. McCarthy & W. Crichlow (Eds.). *Race, identity and representation in education* (pp. 24–39). New York: Routledge.

Bigelow, B., & Patterson, B. (2002). *Rethinking globalization: Teaching for justice in an unjust world*. Milwaukee, WI: Rethinking Schools Press.

Block, S. (2014). Retrieved from https://www.broadbentinstitute.ca/en/blog/sheila-block-reducing-labour-market-inequality-canada-three-steps-time

Boyden, J. (2013). Retrieved from http://www.cbc.ca/thesundayedition/shows/2013/01/13/former-south-african-ambassador-on-aboriginals-in-canada/

Braverman, H. (1998). *Labor and monopoly capital: The degradation of work in the twentieth century*. New York: Monthly Review Press.

Brock, D. (2003). *Making normal: Social regulation in Canada*. Toronto: Thomson Nelson.

Camfield, D. (2011). *Canadian labour in crisis*. Toronto: Fernwood Publishing.

Canadian Press. (2014). Retrieved from http://www.thestar.com/news/canada/2014/04/07/bc_mcdonalds_franchise_investigated_over_temporary_foreign_workers.html

Chomsky, N., & Herman, E. (1988). *Manufacturing consent: The political economy of the mass media*. New York: Pantheon Books.

Cobb, C., & Diaz, P. (2009). *Why global poverty? Think again*. New York: Robert Schalkenbach Foundation.

Galeano, E. (2000). *Upside down: A primer for the looking-glass world*. New York: Henry Holt and Company.

Goar, C. (2014). Retrieved from http://www.thestar.com/opinion/commentary/2014/04/11/harper_government_drives_up_youth_unemployment_goar.html#

Goodman, A. (2014). Retrieved from http://www.truthdig.com/report/item/congress_to_the_unemployed_eat_confetti_20131231

Henry, F., & Tator, C. (2002). *Discourses of domination: Racial bias in the Canadian English language press.* Toronto: University of Toronto Press.

Hessessy, T. (2014). Retrieved from http://www.ldlc.on.ca/uploads/2/7/8/8/2788943/disaster_in_the_making.pdf

Huff, D. (1993). Retrieved from http://www.questia.com/library/journal/1G1-14108114/phantom-welfare-public-relief-for-corporate-america

Illich, I. (1968). Retrieved from http://civicreflection.org/resources/library/browse/to-hell-with-good-intentions

Ivanova, I. (2014). Retrieved from http://rabble.ca/blogs/bloggers/policynote/2014/04/temporary-foreign-worker-program-takes-jobs-away-canadians

Klein, N. (1999). *No logo*. Toronto: Knopf Canada.

Klein, N. (2007). *The shock doctrine: The rise of disaster capitalism*. Toronto: Knopf Canada.

Lorber, J. (1994). Night to his day: The social construction of gender. In *Paradoxes of gender*. New Haven, CT: Yale University Press.

Lyons, K., & Hanley, J. (2012). Gap year volunteer tourism: Myths of global citizenship? *Annals of Tourism Research, 39*(1), 361–78.

MacDonald. D. (2014). Retrieved from https://www.policyalternatives.ca/publications/reports/corporate-income-taxes-profit-and-employment-performance-canadas-largest-compa

Macridis, R., & Hulliung, M. (1996). *Contemporary political ideologies.* New York: HarperCollins.

Make Poverty History. (2014). Retrieved from http://www.makepovertyhistory.ca/learn/issues/end-poverty-in-canada

McChesney, R. M., & Nichols, J. (2002). *Our media, not theirs: The democratic struggle against corporate media.* New York: Seven Stories Press.

McQuaig, L., & Brooks, N. (2012). *Billionaires' ball: Gluttony and hubris in an age of epic inequality.* Boston: Beacon Press.

Mills, A., & Simmons. T. (1999). *Reading organization theory: A critical approach to the study of organizational behaviour and structure.* Toronto: Grammond Press.

Mooney, L., Knox, D., Schacht, C., & Nelson, A. (2004). *Understanding social problems.* Toronto: Thomson Nelson.

Moore, B. (1984). School systems perpetuate racism and don't know it—Can we deal with it? In *Multiculturalism, racism and the school system.* Toronto: Canadian Education Association (CAE).

Moore, B. (1988, April 20). Historical perspectives on anti-racist education, Keynote presentation for Curriculum for the Changing Community Programme d' etudes pour une collectivite en evolution. Sponsored by the Ontario Race Relations Directorate, The Ontario Ministry of Education and the Carleton Board of Education.

Mullaly, B. (2010). *Challenging oppression and confronting privilege.* Toronto: Oxford University Press.

Nelson, J. (2012). Retrieved from https://www.policyalternatives.ca/publications/monitor/big-oils-pandora-box

Nestle, S., & Kanee, M. (2008). *Diversity and human rights in the work environment: A qualitative research study of diversity and human rights in the workplace.* Toronto: Mount Sinai Hospital.

O'Brien, S., & Szeman, I. (2010). *Popular culture: A user's guide.* Toronto: Nelson Education Ltd.

Olive, D. (2014). Retrieved from http://www.thestar.com/business/2014/04/04/can_politics_force_down_excessive_ceo_pay_olive.html

Parenti, M. (2014). Retrieved from http://www.permanentculturenow.com/socialism-for-the-rich-and-capitalism-for-the-poor/

Paul, R., & Elder, L. (2006). *Critical thinking: Learn the tools the best thinkers use.* Toronto: Prentice-Hall.

Ritzer, G. (1995). *The McDonaldization of society.* Newbury Park, CA: Pine Forge Press.

Siddiqui, H. (2013). Retrieved from http://www.thestar.com/opinion/2013/12/11/canadians_did_more_than_anyone_for_south_africa_siddiqui.html

Salih, S. (2002). *Judith Butler*. New York: Routledge Taylor & Francis.

Saul, J. (1986). *The crisis in South Africa*. New York: Monthly Review Press.

Smith, D. (2004). Despoiling professional autonomy: A women's perspective. *Inside corporate U: Women in the academy speak out*. Toronto: Sumach Press.

Spellman, E. (1988). *Inessential woman: Problems of exclusion in feminist thought*. Boston: Beacon Press.

Stanford, J. (2014). Retrieved from http://rabble.ca/print/columnists/2014/04/three-key-moments-canadas-neoliberal-transformation

Storey, J. (2009). *Cultural theory and popular culture: An introduction*. Toronto: Pearson Longman.

Swift, J., Davies, J. M., Clark, R. G., & Czerny, M. (2003). *Getting started on social analysis in Canada*. Toronto: Between the Lines Publishing.

Truscello, M. (2011). *Capitalism is the crisis* [Documentary]. Circle-Eh Pictures.

Wallerstein, I. (1989). *The modern world-system*. San Diego: Academic Press.

Williams, J. (2004). *50 Facts that should change the world*. New York: The Disinformation Company Ltd.

World Wildlife Fund. (2014). Retrieved from http://wwf.panda.org/about_our_earth/biodiversity/biodiversity/

Yalnizyan, A. (2010). Retrieved from https://www.policyalternatives.ca/sites/default/files/uploads/publications/reports/docs/Poverty%20Post%20Recession.pdf

Chapter 4

Applying Concepts and Frameworks of Social Analysis

Chet Singh

LEARNING OUTCOMES

LO-1 Examine dominant ideologies and their origins

LO-2 Compare and contrast dominant and transformative ideologies

LO-3 Review dominant discourses that feed into inequality and other social problems

LO-4 Discuss and assess two frameworks of social analysis

LO-5 Apply a social analysis framework to examine a social issue

Ideology touches every aspect of life and shows up in our words, actions, and practices.... Because ideology structures our thoughts and interpretations of reality, it typically operates often beneath our conscious awareness...it shapes what seems "natural," and it makes what we think and do "right."

—*Eisenberg quoted in Allen, 2011*

INTRODUCTION

To conduct social analysis we need to understand the concept of ideology and identify ideological frameworks. Ideologies refer to a complex set of beliefs, ideas, perceptions, and assumptions that provide members of a society or social group with an understanding and explanation of the community, society, or larger world (Brock, 2003). Ideologies offer a vision of how an ideal society should be organized, what values should be embraced, how we ought to treat each other, and how we should live together as a society (Adams, 2001). In this section we will consider three of the dominant ideologies in the Western world: liberalism, conservatism, and socialism. We will also consider transformative ideologies that attempt to challenge dominant structures of power and exploitation. Finally, we will conclude by looking at two models of social analysis: the triangle model and the self-reflective model.

IDEOLOGY

Ideologies exist in relation to each other. They compete to establish their vision of the best way to organize society, how we should understand societal problems, and how these should be addressed (see Figure 4.1 on page 64). Therefore power relations are central to our understanding of how certain ideologies become dominant and how discourses are used to establish **norms** (socially accepted attitudes, values, and beliefs) that shape what people think, believe, and value about the way society is organized. Various ideologies frame the source of problems according to the interests they serve. Dominant ideologies in conjunction with dominant discourses[1] (how we understand topics or issues and acceptable ways of talking about them) function to legitimize social structures and what we can and cannot say about problems in society. As a result, many people adopt dominant ideological perspectives as commonsense explanations for social problems without being aware of the fact that they are subscribing to a particular ideology, belief, perception, or assumption (O'Brien & Szeman, 2010).

Political and social ideologies are dynamic because they mobilize people to change the structure of society and preserve their interests or what they perceive to be their interests. Since any ideology is vulnerable to challenge, discourses are continually used by such societal institutions as media, schools, and government to reinforce dominant ideologies. Oppressive ideologies are inherently unstable and can be challenged by various groups. These may include groups based on class interests, oppressed social identity groups (e.g., women, **racialized** people, LGBTQ individuals, poor or working class), political coalitions, and religious interests. These groups may question the status quo and attempt to reform it, develop an alternative ideology, attempt to change existing social, political, and economic arrangements, or propose ideas for restructuring society (Storey, 2009).

There are many different types of ideologies. Political ideologies are generally based on concepts of individualism versus collectivism, the size and the role of the state, and the type of economic system. Other ideologies are associated with social and environmental causes (green movement/ecologism), or social identity/human rights-based causes (anti-Apartheid, indigenous sovereignty, anti-racism, feminism, LGBTQ, disability, etc.). For our purposes we will refer to these ideologies as transformative ideologies. Still, other ideologies support nationalist and

norms
Social expectations about attitudes, values, and beliefs.

racialized
The process of creating, preserving, and communicating a system of dominance based on race through agencies of socialization and cultural transmission such as the mass media, schools, religious doctrines, symbols, and images.

---

1. A system of knowledge that limits what we can think and say, who we can be, what we can do, and what can be done to us. The medical model is a discourse, law and order is a discourse, and the accepted theories for various academic disciplines and programs are discourses.

white supremacy

A belief that white people are superior to all other races. This was reinforced through government policy, legislation and laws, and academic and cultural production to glorify whiteness and malign colour.

patriarchy

The systematic privilege and entitlements conferred upon men over women and children through social, economic, and political control.

supremacist causes based on race, ethnicity, gender, religion, and sexual orientation (**white supremacy**, Apartheid, Nazism, **patriarchy**, heterosexism, casteism, ethnicism, and religious fundamentalism).

An awareness of how various ideologies shape our thinking is an essential component of self-reflection and social analysis. First, by examining various ideologies, particularly dominant ideologies, we can identify the values or biases that each promotes. Second, we can disentangle different and conflicting arguments about social problems by understanding how various ideologies shape our understanding of the social world. Third, understanding ideologies helps us to connect our everyday experiences with social, economic, and political processes. We can then question our everyday practices and our experiences. Fourth, once we recognize the biases, ideologies, and assumptions, we can anticipate how others may respond to issues. And finally, being aware of how we interpret social phenomena and social problems can help us to make the connections between ideology and social action (Singh, 2008).

THE ORIGINS OF DOMINANT POLITICAL IDEOLOGIES

The three dominant political ideologies in the Western world over the last three centuries are liberalism, conservatism, and socialism (Wallerstein, 1989). Western political ideologies, and in particular classic liberalism, were all influenced by the Enlightenment,[2] a major intellectual movement in the late 17th century Europe. Classic liberalism as an Enlightenment ideology stood for *free* enterprise, *free* trade, *free* competition, individual freedom and liberty, which was enhanced by equality of rights, constitutional governments, rule of law, and toleration (Adams, 2001). However, these ideas of equality existed in a state of inequality because they did not apply to everyone. Historically liberalism as it evolved under capitalism[3] and imperialism[4] promoted systems of class inequality, patriarchy,[5] and white supremacy.[6]

---

2. An intellectual and scientific movement in Europe that promoted ideas of rational and scientific thought applied to religious, social, and political issues.
3. A recent world economic system based on the ownership of private property where a few own or control the means of production and accumulate capital from the sale of products for profit. Most workers experience exploitation because the value of wages remains below the value of goods produced or services provided. Increasingly, capital is now accumulated through financial speculation. More recently, this led to the economic crisis in the U.S. where banks that created the problem were bailed out with taxpayers' money but those with mortgages involved in this speculative scheme lost their houses.
4. Rule or authority over other national territories through control of their economics, resources, military, and political structures.
5. The systematic privilege and entitlements conferred upon men over women and children through social, economic, and political control. Patriarchy is always present regardless of how men act upon that privilege and whether or not they have privilege over other areas of their lives.
6. An internalized belief developed during colonialism that white people are superior to all other races. This was reinforced through government policy, legislation and laws, and academic and cultural production to glorify whiteness and malign colour. This is similar to casteism in India.

Classic conservatism arose as a rejection of Enlightenment thinking and its ideas regarding the extension of democracy and rights. Its early proponents also had a distrust of *reason* and the appropriateness of using it as a solution for social problems. Religion plays an important role for conservatives because of its deep respect for authority. One of its primary objectives was to reassert the rule of the church and monarchist state as rightful structures to maintain order in a society that conservatives claimed was ordained by God (Adams, 2001). For classic conservatives, individual liberty is more important than equality while hierarchy and stratification (different levels of caste, class, privilege, or status) are the natural order of a functioning society. Conservatives have a deep respect for tradition and embrace change only if it is gradual. As an elitist ideology, conservatives believe that the masses are inferior, some individuals and groups are superior and therefore better suited to be rulers and leaders. For classic conservatives, a stratified and unequal society is natural and ordered and therefore works in the interests of all classes (Mullaly, 2010).

European socialism was consolidated in the 19th century as a reaction to the social and economic impacts of industrialization and the evolving capitalist economic system. Socialism's egalitarian (equal rights and distribution of resources) ethic is diametrically opposed to that of private ownership and private profit because of the inequalities that result from the free market system's competitive individualism favouring those that are already wealthy and privileged.

Socialists believe that we need to understand authoritarian power in order to transform oppressive conditions. To do this, social structures, social relationships, and social change must be understood in their historical, political, and economic context. This means that the material conditions that create wealth and poverty must be examined as historical formations created by powerful elites in their interest. Socialists point out that the capitalist system is based on exploitation since it relies on an unfair labour exchange where the owners of resources profit at the expense of the labouring classes. The capitalist system, they argue, has to be replaced with a system based on cooperation, harmony, and justice. This means common ownership of the economy, no private property, no class system, and equality of wealth, power, and opportunity. Everyone would work for the common good of the community based on the principle of each according to their ability and each according to their need. Its main philosopher Karl Marx proposed that in such a system human nature would not be distorted by poverty or greed and a caring and sharing human nature would flourish (Adams, 2001).

Socialist ideas were influential in shaping Western economic policies by challenging the excesses of class inequality until the rise of **neoliberalism** in the 1970s. Countries that adopt aspects of socialism such as Medicare and social welfare are referred to as social democratic countries. With the dominance of neoconserva-

neoliberalism
An ideology which is premised on the right of individuals to compete in the capitalist marketplace to acquire consumer goods and wealth. It advocates that human well-being is best achieved by creating conditions for entrepreneurial freedoms and skills within a system of free markets and free trade supported by the state (Harvey, 2005).

tive and neoliberal discourses, socialism is perceived negatively.[7] A number of factors account for this. Socialist governments, while achieving some of the ideology's egalitarian aims, have engaged in totalitarian practices such as political repression to maintain governmental power (this was especially true during the Cold War when the Soviet Union and the United States vied for global dominance). Also, since socialist countries generate relatively little wealth for the majority of the population, life tends to be barely above minimum living conditions. Additionally, centralized state control of the political and economic systems tends to reproduce a hierarchy that is undemocratic and places a great deal of trust in political leaders that they will act in the best interests of all members of the society. And finally, socialism has embraced industrial models that are as destructive to the environment as those of capitalism (Adams, 2001; Mullaly, 2010). The overt restriction of individual liberty in socialist countries makes the hegemonic control (power through consent) in capitalist liberal democracies appear mild in comparison.

DOMINANT POLITICAL IDEOLOGIES TODAY

neoconservativism
An ideology that subscribes to similar economic policies as neoliberalism, but uses divisive politics such as racism, sexism, homophobia, the War on Terror, Islamophobia, and immigration to enlist support for its economic vision and social policies (Smith, 2004).

social structures
The network of social relationships created among people when they interact with each other, within societal institutions according to their statuses in that society.

The dominant political ideologies today are neoliberalism and **neoconservatism**. These ideologies are offshoots of classical liberalism and conservatism. Neoliberalism and neoconservatism evolved in the United States and Western Europe. They emerged in the 1970s as an attack on the idea of collectivism, an ethic that considers the collective good of society by developing policies to address inequality and alleviate suffering caused by **social structures.**

Neoliberalism

Neoliberalism promotes individualism and free enterprise ideas and a limited role for government. Neoliberalism's prescription for promoting human dignity and individual freedom is premised on the right of individuals to compete in the capitalist marketplace to acquire consumer goods and wealth. It advocates that human well-being is best achieved by creating conditions for entrepreneurial freedoms and skills within a system of free markets[8] and free trade[9] supported by the state (Harvey, 2005).

A major premise of neoliberalism is that equality of opportunity exists for any individual to pursue; therefore "success" or "failure" largely depends on the indi-

---

7. Americans that stand to benefit from universal healthcare reject it on the grounds that it is a socialist idea.
8. An idea that assumes capital must be allowed unregulated access to national and global markets in order to be competitive. A drawback is that corporations and the wealthy can hide money to avoid taxes.
9. The theory that deregulating markets (eliminating trade barriers, financial controls, and devaluing local currencies), cutting public expenditure (education, health, welfare, etc.) will make exports and hence the economy more competitive.

vidual. Poverty is considered an individual problem, but also points to the failure or ineffectiveness of public institutions. Neoliberals assume that all individuals compete on a level playing field ignoring the structural, economic, and social advantages that privileged groups have accumulated historically. Also concealed are the historic circumstances that created global structural inequality and systemic discrimination against minoritized groups (inquisition, enclosures, colonization,[10] enslavement, and neocolonialism).

Neoliberalism emerged in the 1970s as a **backlash**[11] against liberal social policies and programs. They were created in response to the Great Depression of the 1940s and union demands for fair wages and safe working conditions in the early part of the 1900s in Canada, U.S., Britain, and other Western capitalist economies. These social programs provided government relief for the middle and working classes displaced by the capitalist economic system.[12] These measures also provided government money to stimulate the economy. The financial sector, which was implicated in these problems, was reformed in an attempt to avoid another depression by adding government oversight to the increasingly powerful corporate and financial sector.[13] However, these programs were viewed as a class compromise between the rich and the poor—economic growth through free trade and worker protection as well as social welfare for the underemployed and tax entitlements for the middle classes[14] (Apple, 1993). The wealthiest in society (or "the 1%") made a concerted effort to devise strategies to ensure greater wealth accumulation by influencing government policy and society's perceptions of the working class, and by securing greater control over financial markets.

The central idea promoted to win over public opinion and shape economic discourse was that prosperity for all would only occur when the wealthy and corporations were free from government constraints. First, this required government to deregulate public protection measures such as food safety, pharmaceuticals,

backlash

An adverse reaction to a social ideology that becomes influential such as feminism, or to measures intended to address social problems such as systemic discrimination.

---

10. The conquest and control of other peoples' lands and resources through genocide enslavement and in some instances settlement. Canada being such an example.

11. An adverse reaction, usually from privileged social groups, to a social ideology that becomes influential such as feminism, or to measures intended to address social problems such as systemic discrimination (e.g., employment equity).

12. Historians generally ignore the fact that, at the time, whites were the primary beneficiaries of measures such as preferential home ownership loans as African American and other racialized groups were excluded. Hence, today working class whites have more savings and equity than middle class blacks. See http://blackamericaweb.com/2013/02/13/the-history-of-racial-economic-inequality-part-2-the-new-deal-the-american-dream/

13. http://www.econexus.info/publication/corporations-are-not-human-so-why-should-they-have-human-rights
http://www.umass.edu/legal/derrico/corporateperson.html

14. Most people don't recognize that the financial break that middle classes get through RRSPs and subsidized higher education are public subsidies similar to welfare, however, they are not stigmatized like welfare and viewed instead as entitlements.

environmental protection, genetically modified foods, etc., and replace these with corporate self-regulation. Second, the size of government needed to be reduced. This meant less bureaucracy and, consequently, cutbacks or elimination of social, cultural, human rights, and environmental protection programs. Third, public services such as healthcare, education, social services, prisons, public land, water, etc., had to be privatized (run by corporations). Fourth, taxes for the wealthy had to be cut (Klein, 2007).

Neoliberal discourses were highly successful. Funding to public institutions has been cut drastically and governments have adopted business models for essential public services. Increasingly, aspects of healthcare and higher education have been privatized. Public institutions such as colleges and universities are no longer accessible and tuition for high-paying professional careers has been deregulated (e.g., MBA, legal, and medical degrees) making these programs out of reach for many. Additionally, critics point out that these institutions have been forced to seek out corporate sponsors, which has eroded academic freedom. This is most evident in areas of scientific research sponsored by corporations where research harmful to the public but profitable to the corporations is withheld from publication.[15] Critics claim that neoliberals have undermined structures for democratic decision-making since corporate lobbyists can influence government policy that undermines the welfare of citizens.

The impact after four decades of neoliberalism demonstrates that these policies have achieved their objectives. Governments have been reduced in scope and size, the wealthy have benefited through the deregulation or "freeing up" of financial markets, and many accept this ideology's societal vision as inevitable even though they have been negatively affected.

Neoconservatism

Neoconservatism's primary difference from neoliberalism is its embrace of traditional and fundamentalist social, moral, and religious values. While this ideology subscribes to similar economic policies as neoliberalism, it uses divisive politics such as racism, sexism, homophobia, the War on Terror, Islamophobia, and immigration to enlist support for its economic vision and social policies (Smith, 2004).

Apple (1993) claims that the seeds of neoconservatism were sown in North America and Britain in response to social upheavals such as human rights and anti-colonial struggles in the 1960s and 1970s.[16] These national liberation movements

---

15. For example see the case of Nancy Olivieri vs. pharmaceutical corporation Apotex, at http://www.academia.edu/4004495/Beyond_Academic_Freedom_Canadian_Neoliberal_Universities_in_the_Global_Context; http://dfa.ns.ca/wp-content/uploads/2013/02/OCUFA-Joel-Westheimer.pdf

16. For example, the economic crisis, Watergate, the Civil Rights Movement, Black Power, the Womens' Movement, Red Power, rainbow coalitions, and anti-colonial movements.

challenged neocolonial policies and corporate control of economic resources that perpetuated global inequality. The gradual success of these struggles to advance equity and human rights in North American and European societies was viewed by conservative elites as a breakdown of the moral order of society. Public policy to address discrimination and inequality, such as human rights policies, positive space campaigns, and employment equity, was characterized as political correctness, politicizing education, and attacking the traditions of Western civilization (D'Souza, 1991).

To counter these collectivist social policy developments, a number of neoconservative think tanks were established. They devised strategies to promote the idea that social and economic problems in society were caused by socialist-inspired interventions in public services such as accessible education, welfare, and Medicare. Neoconservatives claimed that these programs were not only the cause of economic problems, they were a drain on the public purse and a slippery slope to socialism (Smith, 2004).

Advocates of neoconservatism argue that "political correctness" is radically altering traditional notions of community, nation, and family. They view multiculturalism, sexual harassment, and human rights policy initiatives as a victim's revolution that has produced a "grievance industry." According to neoconservatives, allegations of racism, sexism, and other forms of discrimination are invented to hide the inability of women and minorities to keep up with college-level work and get academic research grants (Fekete, 1995).

Neoconservatives use race, sexual orientation, and other marginalized social identities as a wedge to gain public consensus on their economic and social vision. They advance a *we* versus *they* status. We are the *real* British, Canadians, and Americans and they are the *other*. We are law-abiding, hard-working, and homogeneous (similar, alike). They are lazy, immoral, heterogeneous (dissimilar, unlike), and multicultural. These distinctions distance the racialized,[17] LGBTQ, and other groups from the community of *regular* individuals. Neoconservatives argue that these *others* are making unreasonable demands. They are taking over and getting something for nothing. Neoconservatives also believe that human rights and equity policies are supporting the *others* and sapping our way of life, draining most of our economic resources, and creating "reverse discrimination" and government control of our lives. In essence, the neoconservative framing of social problems suggests that those that have historically faced discrimination and oppression are now oppressing the *real* Americans and Canadians (Apple, 1993; King & Singh, 1991).

---

17. Racialization is the framework of interpretation and meaning for racial thought in society. It creates and preserves a system of dominance based on race and is communicated and reproduced through means such as the mass media, schools and universities, religious doctrines, symbols and images, art, music, and literature.

Neoconservative discourses were highly successful as people bought into the assumption that social and economic problems in society were the result of collectivist (social democratic) government interventions in education, welfare, health, and programs aimed at increasing opportunities for marginalized groups. These were deemed as expensive and not the role of government. The role of government is to facilitate freedom, which is equated with commercial enterprise not democratic initiatives and collectivist ideals. As neoconservatism became a dominant discourse, there was a backlash against collectivist policies. By the early 1990s government dismantled legislation enacted to bring about equity and structural change such as labour standards, collective bargaining for workers, harassment and discrimination policies, fair employment practices such as employment equity, and environmental protection (Singh, 2004).

Transformative Ideologies and Critical Discourses

> If you've come here to help me, you're wasting your time. But if you've come because your liberation is bound up with mine, then let us work together.
>
> —Australian Aboriginal Elder Lilla Watson attributes this to a women's collective she belongs to in response to academic literature that cites her as the original source of this idea/quote

Transformative ideologies are those that seek to transform systems that create, legitimize, and perpetuate inequalities, injustices, and ecological destruction. Many transformative ideologies have their roots in movements for equity arising out of decolonization, gender equity, and political movements based on class. It is important to recognize that transformative ideologies are not homogeneous and tensions exist within them.[18]

Socialist ideology is considered transformative since it seeks to *change* the structural inequalities of capitalist societies that legitimize the interests of the wealthy and corporations at the expense of the working class. Feminist ideologies challenge patriarchal ideologies, which normalize male domination over women. Feminists point out that male dominance is built into social and institutional practices and experiences. For instance, women often have to do a double shift in the home and at paid employment, are disproportionately the victims of male violence, and are among the poorest social groups in society. Furthermore, women still experience occupational segregation (e.g., daycare workers) in certain sectors, earn less than

---

18. Some emphasize certain forms of oppression while minimizing others. For instance, third wave feminist perspectives have critiqued first and second wave feminist perspectives for the exclusion of racialized and queer perspectives. Similarly, there are anti-racist perspectives that minimize queer and gendered perspectives. And there are feminists and anti-racists that embrace liberal and socialist political ideologies.

male counterparts, receive fewer benefits and pensions, and continue to be under-represented in leadership positions in government and other institutions. Queer theories critique heteronormativity, a belief system that normalizes heterosexual identity and relationships. These norms are reinforced by many institutions such as religious institutions, legal systems, schools, and mainstream media. The result is social stigmatization and violence among other social problems.

Anti-racist/decolonizing ideologies define social problems as unresolved sovereignty rights of indigenous peoples, corporate exploitation of people and resources, environmental destruction, inequalities of income based on gender, race and status, unequal rights and access, patterns of state and hate violence, displacement, and occupation. The elevation of Western culture, arts, philosophy, and science as the standard that all societies must aspire to continues to privilege Western systems of knowledge (Kulchyski, 2014; Lee, 1991).

Ecological ideologies critique both socialist and capitalist devotion to the destructive concept of "progress," which promotes perpetual economic growth and industrial expansion. They argue that these ideas are materialistic, unsustainable, endorse masculine values (hierarchy, competition, aggressiveness, and assertiveness), promote an anthropocentric view of the world (the earth and all living things are there to be exploited for human purpose), and privilege the economy above moral, social, and artistic values.

Singh (2012) suggests that transformative ideologies present vastly different visions of how the world ought to be organized from the current dominant ideologies and have a number of commonalities. First, they put into a *historical* context how specific oppressive and harmful ideologies evolve as relations of exploitation and domination through conscious human activity unlike neoliberalism and neo-conservatism that take an *ahistorical* approach to social problems and obscure the historical roots of social problems (Lee & Lutz; 2005; Mirchandani & Butler, 2006; Prasad, 2006). Examining the historical origins of social problems such as inequality challenges dominant ideologies and the structures they promote. They also provide counter discourses by revealing models of egalitarian and cooperative societies, e.g., matriarchal societies, which have flourished previously. This creates possibilities for organizing economic, political, and social structures based on principles and values that are cooperative and egalitarian (Bishop, 2005; Smith, 2014).

Second, they provide analytical tools and concepts to examine how domination processes function to manufacture, legitimize, and perpetuate Eurocentric, patriarchal, heterosexist, ableist, and classist ideologies and discourses that maintain the dominant group's power (Allen, 2011; Dei & Calliste, 2000; Fraser, 1989; Johnson, 2006; Smith, 2004).

Third, they document how commonsense discourses of citizenship, democracy, freedom, justice, work, and schooling embedded in social structures and practices mask the operation of power, privilege, exploitation, and minoritization (Abu-

Laban & Gabriel, 2002; Barnes & Nazim, 2004; James, 1996; Razack, 2002; Thobani, 2007; Walcott, 2003; Zine, 2001).

Fourth, transformative approaches examine and document the social, cultural, and psychological impacts of oppressive ideologies on minoritized peoples (Bannerji, 1995; Bishop, 2005; The Chilly Collective, 1995; Edward-Galabuzi, 2006).

And fifth, they articulate alternative visions and strategies for working collaboratively on the liberation of all peoples and the protection of the biosphere (Saiz, 2014).

Figure 4.1 presents the dominant neoliberal and neoconservative ideologies with those of transformative ideologies. It serves the analytical purpose of comparing and contrasting the ideas, attitudes, values, and the vision of society each proposes.

FIGURE 4.1 Dominant Ideologies and Their Views of Society and Social Problems

	Neoliberal	**Neoconservative**	**Transformative**
Origins	The 17th century European Enlightenment. Embraces individual rights, freedom, and parliamentary system. Concepts don't apply to the colonized who are exploited, eliminated, and/or assimilated. European wealth and power derived from stolen and appropriated resources and territories. Social ideologies of white supremacy and patriarchy justify enslavement and destruction of indigenous cultural, social, and economic systems. Promotes values of competition, individualism, and consumerism. These usher in industrialization and the capitalist economic system.	Arose as a rejection of Enlightenment concepts of freedom, equality, and individual rights for European peasants and serfs. An elitist ideology that believes in tradition, supports the return to rule by aristocracy and the Church. Asserts moral superiority of aristocracy and other elites. Supports coercion or the threat of it. Supports discriminatory social ideologies based on race, class, and gender. Supports hierarchy, religion, and social Darwinism—some are naturally "more fit" to rule than others.	Archeological evidence indicates that early human societies were egalitarian and many were matriarchal. Challenges oppressive conditions of white supremacist, capitalist, patriarchal ideologies. Includes collectivist ideologies such as socialism, anarchism, feminism, anti-racism, queer activism, decolonization/Indigenous sovereignty movements, and environmentalism. Proposes alternatives based on equity, sustainability, cooperative structures, and non-hierarchical arrangements.

Source: Bishop, 2004; Korten, 2001; Mullaly, 2012; Singh, 2004, 2012.

	Neoliberal	Neoconservative	Transformative
Assumptions About Human Nature	People can learn reason. People are moral, rational, competitive, cooperative. Recognizes class but claims people succeed and can transcend class based on merit. Inequality is caused by individual deficiency, pathology, lack of education, and other circumstances.	People are flawed. People are competitive, individualistic, acquisitive, and motivated by financial gain and power over others. Greed is good, it is the engine of "progress." Hierarchy is based on a God-given right. Some groups are naturally superior; others inferior. European civilization is superior to others. Men are superior to women. LGBTQ is unnatural. Whites are superior to other "races." The poor are inferior.	People are basically good. People acquire values through socialization. We are not born selfish, competitive, racist, sexist, or homophobic. Historically, humans created egalitarian, cooperative, and collective societies. Many indigenous societies were based on cooperative and egalitarian structures and beliefs.
Views of Human Rights	Assumes that we are living in a post-discriminatory age, because individuals are protected by legislation. Does not acknowledge historical and structural inequities that advantage some and disadvantage others. Believes that we all start from a level playing field and the welfare state compensates victims of capitalism. Believes in tolerance of all differences. Supports multiculturalism.	There are no social problems, only individual problems caused by individuals who do not look after themselves. Poverty is good because it teaches discipline and provides incentives. Individual liberty is more important than equality. Human rights policies are unjust, "politically correct," censor free speech, and promote "reverse discrimination" against white males. Does not support same-sex relationships.	Inequality and discrimination are historically embedded in social structures and obscured by dominant ideologies and discourses. Exploitation is gendered and racialized and utilizes child labour. Imperialist governments have engaged in cultural and physical genocide and suppressed the collective rights of indigenous peoples by refusing to settle land claims and sovereignty rights. The state must promote education and dismantle oppression and inequality.

	Neoliberal	Neoconservative	Transformative
Role of the State/ Organization of Society	To promote fair competition among stakeholders through regulation and state intervention. To reduce the worst excesses of capitalism. Culture and society dedicated to idea of trade and consumption as the ultimate source of well-being. Corporatist capitalism and trickle-down theory: we all benefit when the rich get richer. Everything should be commodified and privatized, size and role of government should be reduced, economic oversights deregulated along with environmental protection.	A necessary evil that maintains law and order for the management of trade and commerce. Anything beyond this is a threat to liberty. Assures conformity of subordinate groups to interest of dominant groups. Institutions support and serve the dominant ideological base—hierarchical class-based consumerism and imperialism. Similar economic policy to neoliberalism. Opening up of markets globally, people will be displaced, a necessary aspect of wealth accumulation/progress. Facilitates the dismantling of trade unions, labour protections, Keeps price of labour low with programs such as seasonal and temporary foreign workers.	State should be a participatory democracy that facilitates social justice, equity, environmental protection, and equitable distribution of society's resources. Corporations need to be regulated in the public interest. They are now the dominant governance institution on the planet. They have too much influence on the nation state and collude with government against citizens interests. The global north must end neocolonial and imperial policies. Economic growth must be balanced with social priorities and environmental protection.
Social Change	Social reforms that benefit individuals are OK as long as they keep the system intact. Does not support transformative change. All social and economic adjustments or change must be consistent with the existing system. Conflict must be managed. Educational programs to change discriminatory attitudes promoted over structural change. Focus on charity and voluntarism: "helping" the "disadvantaged." "Learning" about the disadvantaged and getting to "know" and making connections with the disadvantaged.	Keep the system the way it is for the common good, it is natural and right, inequality has always existed. There are those that are superior and those that are inferior. We must take or be taken, defend what we have, the world is competitive. Does not acknowledge systemic issues and does not believe that the state should be involved in social justice or other initiatives aimed at creating equity or justice. Individuals must look after themselves. Any change should be slow and evolutionary.	Change must involve everyone. Conflicts are caused by unjust systems but can be a creative source for transformation. Must work collaboratively to seek social structures that are just, equitable, and fair. We need to work on our own liberation while being an ally with other forms of oppression. We need to become informed about different oppressions and how they are connected. Understand how oppression came about, how it is maintained within individuals, recreated within institutions and social structures. Capitalism and dominant oppressive ideologies adapt to challenges and reformulate to maintain legitimacy. Structural change is very difficult to achieve.

	Neoliberal	Neoconservative	Transformative
Environment	The earth, animals, plants are resources to be commercially exploited. They only have value when they are commodified. The only way to protect nature is to give it a commercial value. Perpetual growth is a necessary and natural form of progress and takes precedence over environmental protection.	Control over nature, mastery over nature ordained to "man" by "god." Nature is harsh and cruel. We are separate from and in conflict with nature. Against environmental protection, regulation, species protection, and for opening up protected areas such as national parks for resource development.	The biosphere has intrinsic value. Some view nature as sacred. Humans are not separate from or superior to nature. Nature gives life and is alive. Destroying nature will ultimately lead to our demise.

THEMES THAT DOMINANT DISCOURSES USE TO LEGITIMIZE INEQUALITY AND INJUSTICE

> Those who have studied worldwide liberation struggles know that the manipulation of information, including propaganda and disinformation, are primary tactics employed in the domination process. Oppressive populations defame, stigmatize, stereotype, and distort the reality of dominated populations. Oppressive populations change the true human record through denial of the very reality of the total human experience, including their own.... The mass media, entertainment, and schools are an integral part of the information process and, in oppressive societies, are a part of the domination process as well.
>
> —Hilliard, 1992*

Powerful groups have the means to promote particular ideologies and circulate discourses that serve their interests and legitimize their authority. Gaining consent from the population through the adoption of these ideas is referred to as hegemony. For example, the capitalist economic system creates vast inequality, a trend that is increasing at a rapid rate (Piketty, 2014). However, most believe that it is the best economic system ever devised, that for the most part it works well for everyone, and we must fight to preserve this production and consumption driven way of life, despite the costs to those who are exploited and the negative environmental impacts. This consensus has been achieved through the production of dominant

---

* Asa Hilliard III, "Why We Must Pluralize the Curriculum," *Educational Leadership*, Dec/Jan, 1992, Pg. 12.

discourses. Dominant ideologies and discourses have been challenged throughout history by transformative ideologies and movements that expose how oppression, inequality, and injustice operate to enrich powerful groups and maintain power. In order to maintain hegemony, discourses must be altered or reinvented to maintain legitimacy and manufacture consent from members of a society. Since common-sense assumptions derived from dominant discourses obscure the root causes of social problems, or label them as individual problems, the task of social change is a formidable task (Chossudovsky, 2003).

By examining the hidden assumptions and arguments within dominant discourses and ideologies, we are able to make sense of the dominant culture's current hegemonic practices and reveal structural causes of problems in society. Our society is structured to create vast economic, social, cultural, and political inequality. However, many believe that "developed" nations—the West, ruling elites, and privileged social groups—deserve their position because they have achieved their status through superior effort and means. The oppressive, exploitative, and discriminatory systems necessary to maintain this position are concealed in the popular discourses of mainstream media and schools. These hegemonic arrangements normalize the status quo and discredit alternative visions of organizing society to benefit all members, such as living cooperatively, and protecting the ecosystem for its intrinsic value for the survival of future generations (Parenti, 1995). Rather, we accept warfare for territorial control and resource acquisition as normal. Discussed below are a number of commonsense assumptions or **hegemonic discourses** that perpetuate inequality and oppression that are useful for the analysis of social problems in the world today.

Inequality, Hierarchy, Warfare, Exploitation, and Domination Are Normal

hegemonic discourse
A way of framing issues that becomes so embedded in a culture that it appears silly to ask "Why?" about their assumptions. Commonsense assumptions predetermine answers, and also influence the questions that can be asked.

Conservatives tend to argue that hierarchy, inequality, and competitive individualism are natural, effective, and efficient ways of organizing society. The assumption is that society is imperfect and flawed. The "superior" should rule and organize society accordingly because the majority are "inferior," dependent, easily manipulated, and psychologically weak (Heywood, 2007). This ideological assumption normalizes inequality as inevitable.

Archaeological evidence and knowledge of human social, cultural, economic, and political systems indicate that dominance, individualism, inequality, and hierarchy were not common to all societies. Historically, humans exercised choice in the design of societal organizations. Early recorded history reveals that many societies were egalitarian, cooperative, and gender equal. The land was held in common and any surplus land supported common purposes such as religion, the arts, crafts, and literacy. Inequality and poverty developed when the aristocracy in various societies appropriated land for their private use and a few became very wealthy and powerful. This template was used by imperial powers throughout history in various parts of the globe (Bishop, 2005). More recently, this occurred during the early stages of capitalism in Europe when common land was made into private property for the aristocracy and emerging commercial elites. This process was also used during colonization where land and local technologies were appropriated, and spiritual, economic, and political structures were destroyed to create cheap labour and dependency on the colonial power. This system has served wealthy groups and imperial nations at the expense of others. Consequently, social and economic structures that create and maintain poverty and inequality globally persist. The United Nations reports that the economic gap between the richest and poorest country during colonization was 3 to 1 in 1820, 35 to 1 in 1950, and 74 to 1 in 1997 (Cobb & Diaz, 2009).

In these historical examples, the oppressed and dominated have always resisted the imposition of individualistic values and hierarchical social organization. We continue to see evidence of this today as many indigenous nations in Canada and elsewhere resist assimilation by dominant economic, political, and social ideologies that are rapidly encroaching on their remaining traditional lands. Canadian mining companies, hydroelectric dams, and oil exploration corporations in the Amazon are among those implicated in these ongoing neocolonial processes. However, the general public is ambivalent, unaware, or unconcerned about these processes because little attention is given to these struggles and the assumption is that eventually these "uncivilized" "remnants of the past" will be absorbed by the inevitable "march of progress" (Davis, 2009).

There Is A Scarcity of Resources and Food For the Global Population

The myth of scarcity deflects attention away from the fact that the richest "1%" of the world owns 32 percent of the wealth (Cobb & Diaz, 2009). According to Bishop (2005), there are enough resources in the world to provide for everyone. However, the myth of scarcity is frequently used to disguise the fact that a large proportion of the world's resources benefit very few people. The world's resources are enough to give everyone a good quality of life. The problem is that the current economic system is structured to benefit the few, which creates and maintains class and global inequality. We in the West consume 80 percent of the world's food and energy resources when we are only 20 percent of the population. Many of us in the West die from diseases related to overconsumption and stress while 24,000 children die from hunger every day. The richest 20 percent of the global population make 86 percent of consumer purchases and create 53 percent of the world's carbon emissions, while the poorest 20 percent make 1.3 percent of consumer purchases and create 3 percent of the world's carbon emissions. One person living in the U.S. consumes a hundred times as much of the world's resources as a citizen in one of the poorest countries and causes one hundred times the environmental damage (Lester, 2005). It is estimated that in the West we throw away 40 percent of our food; in Canada that adds up to $27 billion annually (CBC, 2014).

The myth of scarcity is also a useful social Darwinist[19] discourse that powerful countries (G8) and multinational corporations promote to legitimize international resource wars and unfair international trade structures (IMF, World Bank, and the World Trade Organization). The logic behind this neoconservative view is that only the fittest of the fit will survive in a competitive world.

The "Economy" Is More Important Than Anything Else

Dominant narratives derived from capitalist ideology and Christian religious ideology, fundamental to the colonizing processes, and global economic systems today, suggest that nature has no intrinsic value other than its commercial value. "God gave man dominion over the animals and the earth," therefore the biosphere exists for the needs of humans. In contrast to other worldviews that do not privilege humans over nature, Western worldviews portray nature as a commodity to be exploited for commercial gain. Corporations now have tremendous sway and influence over government policy and stand to benefit most from this. Consequently, perpetual economic growth takes precedence over harmful and environmentally destructive practices (Conally, 2008). When these harmful, destructive, and unsustainable practices are challenged by environmentalists and scientists, discourses are used to discredit and dismiss their claims. This is most apparent with governments such as Canada's that work in concert with the oil corporations and subscribe to these models of economic growth. The Canadian government has put out ads in mainstream media suggesting that environmental problems related to the Keystone pipeline project are exaggerated by activists and radicals who are funded by foreign interests. The Canadian government also ignores their own scientists whose studies reveal that the Alberta tar sands are poisoning the ecological system and affecting the health of indigenous communities that depend on the land. Canada is the only country that signed and ratified the Kyoto protocol and then said it had *no* intention of meeting its targets. The federal government eliminated the only major federal renewable energy program in the country while providing over $1 billion a year in taxpayer dollars (corporate welfare) to subsidize the oil sector. Not surprisingly, Canada has won the Fossil of the Year award five times

---

19. The application of evolutionary biologist Charles Darwin's theory of natural selection to human social life. Economic principles were deemed similar to evolutionary principles and thus only the "fittest of the fit" survive.

(awarded to the country with the worst environmental reputation) (Climate Action Network, 2014).

Corporate economic activity is having a detrimental impact globally but Western countries, as well as India, China, and Japan, have done little to rethink the concept of development. As a result of global industrial fishing practices, only 10 percent of big ocean fish remain, and 85 percent of all oyster reefs have been decimated. Rainforests that once covered 14 percent of the earth's land surface now cover 6 percent. It is estimated that at the current rate of deforestation the last remaining rainforests could be consumed in less than 40 years. Scientists estimate that we lose 50,000 animal, plant, and insect species every year due to rainforest deforestation. For corporations and the government policies that support their economic interests, short-term gain trumps long-term planning and sustainable development for future generations.

The "West" Is Best: Developed, Democratic, and the Apex of Civilization

Critical anti-racist and feminist scholars maintain that the educational system, "experts," mainstream media, and popular culture reinforce ideologies of white supremacist capitalist patriarchy through various discourses and media (hooks, 1994; West, 1993). Textbooks, movies, and television shows, nursery rhymes, social commentary, literature, jokes, pictorial depictions, and other cultural mediums establish Western civilization, competitive individualism, heteronormativity, patriarchal gender roles and norms as the standard that all societies should aspire to. Other knowledge, cultures, and forms of social and political organization are omitted, distorted, or misrepresented through stereotyping and dehumanization.

This misinformation is deemed to be objective and factual because it is backed by the legitimacy of powerful media and academic institutions. One example of this is the misrepresentation of the origins of the Industrial Revolution and its relationship with global inequality and notions of progress and development.

> History is still taught from the perspective of northern Europe as if it were "universal history." In this narrative, the initiative that transformed Europe and made it the centre of the modern world was the industrial revolution. That story distorts the origins of the capitalist order and the world system. Northern European development was preceded by the colonization of Latin America by southern Europe and the transfer of gold, silver and the so-called "precious vegetables," above all sugar cane…. to the north…. This permitted them to accumulate wealth and serve as the basis for…. the industrial revolution. The industrial revolution appears in the guise of universal history as something separate from colonialism. (Cobb & Diaz, 2009)

The Dutch and other European colonizers were particularly barbaric in their exploitation of the countries they colonized. They destroyed the Indonesian textile and ceramic industries, and then took these techniques back to Holland to build a profitable industry. By taking from Asia everything they could they became the world's financial centre before it was transferred to Britain. In the 18th century the British destroyed the Indian textile industry, which was superior to theirs, appropriated the technology, brought it back to Britain, and prevented merchants within the British Empire from importing Indian textiles. These practices destroyed social structures and livelihoods and led to famines. It is estimated that 25 million Africans died during the slave trade, 65 million indigenous peoples were eliminated in the Americas, and 9 million women were killed during the Inquisition (Anderson, 1995; Jones, 2014; Shovel, 2014).

When colonized countries gained political independence, they were left with the debts of the colonizers in violation of international law. This strategy was used to exert control over their natural resources through debt. Another legacy of colonization, designed to make sure that former colonies remain poor and dependent, is the export of raw materials to be processed in Europe and North America and the finished products exported back. Policies of the WTO, IMF, and World Bank help to facilitate the transfer of wealth to the North. Economist Susan George has calculated that the South finances the North $200 billion every year (Cobb & Diaz, 2009).

Misinformation and lack of awareness lead many in the West to conclude that our relative wealth is based on our superior culture, technology, and economic system. The "others" are impoverished because they are backward, uncivilized, undemocratic, and violators of human rights. Furthermore, democracy is equated with parliamentary electoral systems and conflated with the economic system of capitalism. The U.S. has two political parties that firmly support corporate capitalism. There are 67,000 people employed as lobbyists in Washington D.C.—that makes 125 corporate lobbyists for each elected member of Congress (Williams, 2004). The U.S. and their allies have invaded democratically governed countries on numerous occasions to secure geopolitical and resource interests.[20]

Popular culture and academia present the Western way of life as universal and self-evident. Global cultures are evaluated as "modern" based on their similarity to Western consumer capitalism. Furthermore, economic trade liberalization resulting from globalization has had the effect where non-Western societies aspire to the "American way of life," which is regarded as the pinnacle of human achievement. In schools, business texts regard capitalism as a given. They contrast ideal capitalism

---

20. http://academic.evergreen.edu/g/grossmaz/interventions.html http://www.thirdworldtraveler.com/Blum/US_Interventions_WBlumZ.html

against real socialism as implemented by totalitarian states such as the USSR and North Korea[21] (Paul, 2014). In addition, much critical analysis is omitted and other economic systems are not considered worthy of study or examined in relation to capitalism. The growth of monopolies and corporations and government programs such as corporate welfare that contradict capitalist theory is omitted from analysis.

We Are Meritocratic and Democratic: Anyone Can Succeed If They Try Hard Enough

Another discourse that minimizes inequality and discrimination is the myth of meritocracy. This discourse suggests that North American society is based on equal opportunity on a level playing field. The assumption is that all it takes to succeed is hard work—you will be hired based on merit, not your race, class, gender, disability, age, etc. Therefore individuals are responsible for their own success or failure. To some extent this is true, but only for the very few.

This discourse promotes the assumption that wealthy elites deserve their wealth and success, having achieved these entirely through hard work. While this may be true in a small number of cases, unacknowledged is the reality that this wealth has been acquired through exclusion and/or exploitation during colonization, through differential distribution of resources, and differential treatment of minoritized groups. The vast majority of those who are wealthy and successful have advantages that others don't. They enjoy privileged access to jobs, social capital, preferred citizenship, and preferential treatment from societal institutions and structural advantage.

Western societies are highly stratified along lines of race, class, gender, disability, and sexual orientation (Edward-Galabuzi, 2006; Nestel & Kanee, 2008). Gendered, racialized, and disabled minoritized groups are the poorest in society because they supply cheap labour and do most of the dangerous jobs in society. However, discriminatory discourses suggest that "we are providing jobs for them," "they are newcomers and all newcomers have to go through this stage." In the case of gender, women continue to hit the glass ceiling because of stereotypical discourses that marginalize women: "women are not breadwinners," "women have kids, therefore are unreliable," women are not as capable as men, or "women are too emotional." Women continue to earn less than men doing the same job, and only about five of the top Fortune 500 company CEOs are racialized and female (Mullaly, 2010). Yet, most who subscribe to this discourse would consider systems of inequality as undemocratic or unfair.

---

21. This does not represent an endorsement of socialism. As previously mentioned, socialism as practised by states like Russia and China did not live up to its egalitarian ideal. This is largely because of hierarchical and militaristic governmental structures and geopolitical competition with dominant imperial capitalist states like the United States and England.

North Americans Are Living In A Post-Discriminatory Era

Human rights legislation, the denial of privilege, tokenism, and the backlash against equality programs, unions, feminism, and anti-racism are representative of discourses that suggest that we are living in a post-discriminatory society. In reference to race, Henry and Tator (2002) have referred to these contradictions as democratic racism. We may have friends of other "races" at work, but we continue to hold on to stereotypes, do not have deeper connections such as marriage, and may even engage in discriminatory behaviour, all the while thinking we are progressive. Other forms of denial are found in discourses that blame the victim. This belief blames minoritized group members for their own oppression. In North American and other societies women continue to be blamed for male violence such as rape and harassment. Misogynistic ideas that circulate in social structures and in legal institutions often re-victimize women. The first questions asked of women who have been raped is, "what was she wearing?" "what did she do, was she drinking?" and "what kind of woman is she?" This was exactly the inference made at a recent lecture at a Toronto law school by a police sergeant who noted that "women should avoid dressing like sluts in order not to be victimized." Women responded by organizing a "slut" walk to challenge these sexist ideas that continue to persist.

Goodman (2001) claims that dominant discourses have a tendency to entitle people from privileged groups. Because it's not likely to be their experience, therefore it doesn't exist; she refers to this as "willful innocence." Along with a sense of superiority it becomes easy to proclaim that discrimination and oppression does not exist. When claims about discrimination or harassment are made, those raising them are dismissed as politically correct, as having a chip on their shoulder, as oversensitive or as radical trouble makers.

We In the West Are Good and Charitable

"We are the World," "Live Aid," and "World Vision" discourses that promote volunteerism, charity, and foreign aid make those in the West look and feel generous, alleviate symptoms of inequality, which is worthy in itself but does very little to change the structures that create and maintain suffering and oppression. Regardless of our socio-economic positions, those of us in the wealthy corners of the world are afforded the opportunity to feel good about ourselves by giving up a few luxuries to donate or volunteer overseas. Pop stars can sell a lot of records and advance their profiles by getting involved in charitable causes. In fact, charities themselves have become businesses where most of the money goes to advertising and fundraising and a very small percentage goes to the actual cause. The non-profit organization set up by Bono of U2 received £9.6 million (C$17.5 million) in donations in 2008 and handed out only £118,000 (C$215,600) to charity (1.2 percent); £5.1 million (C$9.32 million) went towards paying salaries (Patterson, 2014).

SOCIAL ANALYSIS METHODOLOGIES

Activism is about how we actively change the world.

—Julia (Butterfly) Hill

Throughout the last two chapters we have introduced concepts such as critical thinking, power, ideology, discourse, and hegemony that can be applied to social analysis methodologies. Engaging in social analysis means gathering and sifting through information in order to arrive at an informed opinion about social problems. Social analysis concepts and methods are essential components of the social analysis tool kit; they are not rigid guidelines. They can be applied creatively to analyze specific social problems. Two models of social analysis are described below and one is applied to the documentary film, *Black Gold,* which addresses an aspect of global inequality.

The Triangle Model of Social Analysis

The triangle model (Figure 4.2) is useful for analyzing how hegemony (power through consent) is expressed through the interdependent relationship between dominant ideologies, societal institutions, and individual beliefs and actions.

This model helps us to identify how dominant ideologies frame social problems, how societal structures and institutions reinforce these ideas, and how we adopt influential ideas and discourses as commonsense explanations of social problems. It also reveals how the present arrangement of society relies on our consent and participation as we embrace particular perspectives and support the ruling system through our actions as citizens and consumers.

This triangle model analyzes how particular ideologies rationalize structures of inequality and oppression. It examines how dominant ideologies marginalize and exploit subordinate groups and societies through the omission and misrepresentation of facts, as well as through objectification, stereotyping, and dehumanization. This model also reveals how dominant discourses and language are used to legitimize the position and actions of those that benefit from inequality and exploitation and blame the victims of these structures of domination. It also requires that we examine the social, psychological, physical, and economic impacts of harmful ideologies and discourses.

The triangle model draws a connection between dominant ideologies and institutional structures such as the government, financial institutions, legal systems, corporations, media, schools, and religious organizations. It requires that we examine how particular ideas are normalized through institutional policy and practices that can mask exploitation, inequality, discrimination, and marginalization.

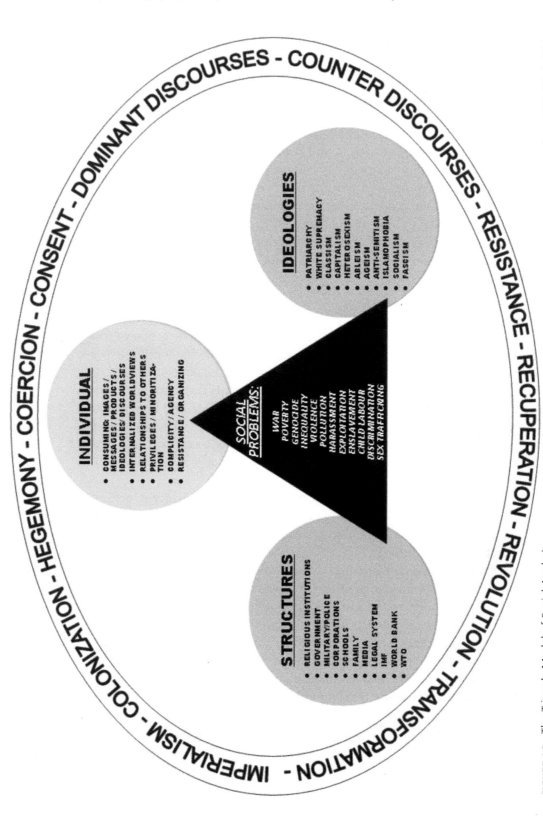

Source: © Chet Singh, 2014.

FIGURE 4.2 The Triangle Model of Social Analysis

Finally, this model examines how dominant ideologies and discourses and social structures influence our individual expressions through the commonsense assumptions we adopt and our everyday behaviours. How do we as individuals participate in these structural and social relationships? What do we know about relationships of domination and exploitation? What don't we know? What are the consequences? How do our beliefs and actions support relations of domination? How can we resist inequality and exploitation through individual or collective agency?

Using Documentary Films for Social Analysis

We will illustrate how to apply the triangle model to a social problem using the documentary film, *Black Gold* (2006), which examines why Western corporations make billions from coffee annually but the small farmers that grow the beans are impoverished.

Documentary films are excellent educational tools for social analysis. The viewer experiences the information on multiple levels. In addition to revealing information that can be empowering, enlightening, or disturbing, films can be visually stunning pieces of art. Since they relay information through storytelling, at the emotional level we can see the human and ecological impact of issues and experience the subject's anger, joy, or insight. At the cognitive level we get the filmmaker's ideological point of view. Documentary films present the filmmaker's version of "the truth." Therefore, it is impossible to completely arrive at an accurate portrayal of events, people, organizations, or issues. As we view documentaries on various issues consider the following critical thinking questions as you gather information to analyze the social problems and solutions presented by the filmmakers:

- What is the dominant ideology of the film? Who funded the film? How does the film construct ideas about the social problem? Does the filmmaker's ideological bias prevent her/him from considering other perspectives on the issue?
- What information is used to examine the social problem? Has information, relevant voices, or sources been omitted that could alter the perspective of the issue, broaden the issue, make important interconnections with related issues, and alter the conclusions or arguments?
- Who are the experts in the documentary? Who are they affiliated with and what is their background or involvement with the issue? Are they directly affected by the issue or distant observers?
- What social identities do they represent (occupation, political affiliations, race, gender, class, religion, sexual orientation, disability, age). Who benefits from the dominant perspective in the film?
- What techniques are used in the film? Does music or framing attempt to manipulate your emotions or create a negative or positive bias towards

a subject or situation? Is this a perspective that would be represented in mainstream corporate media? What do you know about the social or environmental problem prior to viewing the documentary? How did you come to know about the problem?

- Does it challenge your commonsense assumptions/worldview? How does this cause you to react?

The Coffee Trade: *Black Gold*, 2006

The filmmakers state that *Black Gold* (2006) was made to challenge commonsense assumptions and stereotypical views about poverty in African countries by exposing the collusion among global financial institutions, multinational corporations, and Western governments to extract wealth from the global South. The narrative of the film and how it constructs what we know about the coffee trade is influenced by transformative ideological perspectives.

The documentary examines global financial institutions and government and corporate structures to demonstrate that there is an ongoing neocolonial relationship of power that relies on the participation of Western consumers. Trade regulations designed by the powerful Western institutions enlist our participation as consumers in a system that keeps farmers in developing countries impoverished. The four corporations that control the coffee trade make billions annually while the Ethiopian coffee farmers live in poverty because they are not adequately compensated for producing coffee. In fact, the price they are paid has been steadily dropping for the last 30 years. The filmmakers requested interviews with the corporations that control the coffee trade but all declined to be interviewed so their voices are absent from the film.

Individual

The film's narrative centres on Tadesse Meskele, a man who attempts to save from bankruptcy the 74,000 struggling coffee farmers he has organized into a fair trade cooperative. To make the connection to Western consumers, the film switches between the coffee farming communities in Ethiopia to Western urban scenes as hipsters and others consume coffee as they go about their daily business unaware that they are key participants in a global trading system that is unfair. It creates great wealth for four multinational corporations off the backs of poor farmers who produce coffee, bananas, and other commodities.

The film challenges neoconservative and neoliberal assumptions of global inequality common in stereotypical and decontextualized mainstream media representations that many adopt uncritically. Mainstream movies and news stories present "Third World" subjects and countries as "underdeveloped," impoverished, and helpless. They construct the social problem of inequality in African and "Third

World" countries as primarily arising from the actions of African countries and Africans. At the individual level, the video reminds us that our beliefs and actions have powerful social, economic, and political consequences. The cup of coffee or tea we consume at Starbucks or Tim Hortons is not just a satisfying buzz or a personal expression of brand appeal; it is attached to historical colonial relationships of inequality, embedded in capitalist consumer ideologies, and controlled by political and financial systems of power.

Ideology

After oil, coffee is the most valuable exported commodity from the "Third World." Neoliberal capitalism promotes the idea that inequality is the cost of doing business in a competitive environment. Discriminatory media discourses suggest that global economic and social inequalities are outcomes of a natural hierarchy and the superiority of the West. These discourses are historically embedded in socially constructed discourses such as "First World" and "Third World." The former is the term used to describe the industrialized countries of Europe and North America. Third World characterizes the impoverished countries of Africa, Asia, Pacific Rim, and Latin America, which are also called "underdeveloped." Eurocentric historians and economists frame "underdevelopment" and impoverishment as an original historical condition. It is never framed as rooted in the history of colonization and imperialism. This dominant narrative tells us that these conditions have always existed; "Third World" peoples are always "backward," in need of "civilizing," development, and charity; that this is an ongoing project of the benevolent West.

Historical Context

Parenti (1995) reminds us that Europeans went to these countries to steal and plunder the vast treasures of food, minerals, and other precious resources (gold, silver, furs, spices, timber, sugar, rum, rubber, tobacco, coffee, cotton, copper, tin, iron, ivory, oil, zinc, manganese, mercury, platinum, cobalt, bauxite, aluminum, uranium, etc.). The Third World has always been incredibly rich; it is its peoples that have been and continue to be impoverished by exploitative and oppressive neocolonial practices and **comprador elites**.

comprador elites
The local business class who derive their wealth and status from multinational corporations and links to imperial and/or former colonial empires.

European society in the 15th to 19th centuries was not the epitome of cultural supremacy. Europe led in hangings, murders, other violent crimes; venereal disease, small pox, typhoid, and other plagues; social inequality and poverty; abuse of women and children; famine, slavery, prostitution, piracy, religious massacre, and inquisitional torture (Parenti, 1995). Chossudovsky (2003) proposes that superior firepower, greed, and brutality, not superior culture is what brought Europeans and Euro-North Americans to positions of power and that today is still maintained by force and hegemonic discourses. Social customs of indigenous people in the South were characterized by European "explorers" as humane and less autocratic and repressive than European society at the time. Southern civilizations also had superior technologies in textiles, ceramics, and impressive skills in agriculture, architecture, and medicine. All had philosophical, artistic, and cultural traditions. Nonetheless, these were diverse societies and some had cruel and brutal practices of their own (Parenti, 1995). However, much of the information about colonized societies are omitted or distorted by Western historical and academic discourses, to justify their subjugation.

Structures

As European colonizers gained control over local economic structures, food supply systems were dismantled, local technologies appropriated, and colonial economies were structured to extract wealth for European elites. Coffee was one of the commodities that fuelled European colonial economies by replacing local food production and industries with mono-crops for export to Europe. New markets were created for coffee consumption in Europe and colonies were now made dependent on European manufactured goods (Cobb & Diaz, 2009). This model was repeated with other commodities and contributed to the wealth of Europe and North America and the "underdevelopment" or impoverishment of the global South (O'Brien & Szeman, 2010). Though the nations of the South have gained political independence in theory, they certainly do not have economic independence because they continue to be exploited by former European and Euro-North American imperialist powers. Trade regulations that benefit Western economies and corporations force developing countries to adopt agricultural practices that consign them to continued poverty.

Global neocolonial trade structures and capitalist institutions such as corporations continue to enrich the West and exploit the Third World's' labour and resources through the use of force (invasions and assassinations), and the manipulation of financial and commodity markets (Perkins, 2008). According to economists and African politicians interviewed in the film, trade regulations imposed by global financial institutions such as the World Trade Organization, the International Monetary Fund, and the World Bank facilitate neocolonial exploitation of the South's commodities that perpetuate poverty and inequality. The WTO is controlled by the G8[22] and governs the rules of global trade. It has imposed arbitrary rules that are undemocratic and unfair to the South but benefit the West and its corporations. For instance, European and North American governments are allowed to subsidize their agricultural corporations up to $350 billion with tax dollars. As mentioned previously, this is money given to corporations by government and referred to as

---

22. The G8 is comprised of Canada, France, Germany, Italy, Japan, Russia, United Kingdom, and the United States.

"corporate welfare." These subsidies are designed to drive down the price of commodities since Western corporations are then able to sell their produce below the cost of production. Under these trade rules "Third World" farmers are not allowed subsidies, nor can their governments provide them. African politicians point out that these policies have reduced Africa's share of global trade to 1 percent. What impoverished African countries need is trade—not aid. For example, a 1 percent increase in trade will result in the generation of $70 billion, five times more than the continent receives in aid annually (Francis & Francis, 2006). Third World countries have suffered a 70 percent drop in the price of agricultural exports compared to manufactured exports (Cobb & Diaz, 2009).

Impacts

Another structural reason for the growing inequality is that the coffee industry is now controlled by four western corporations (Proctor & Gamble, Sara Lee, Kraft, and Nestlé) who exercise disproportionate control over the wages received by the growers and the prices paid by consumers. The price of coffee paid to southern farmers has been declining over the last 30 years since the collapse of the International Coffee Agreement. Most of the world's coffee is produced by the poorer Third World countries where 25 million families on small farms and highly exploited low wage workers eke out a living. Most farmers receive $0.14 per kilogram for green beans, which are sold to the large multinationals who control the roasting, marketing, and distribution. Multinationals make $26.40 per kilogram—a 7000 percent increase in profits. In 2003 Starbucks' CEO, Orin Smith, was paid over $38 million—the combined annual income of 36,000 Ethiopian workers (Fridell, 2012).

In Ethiopia many farmers have been forced off the land into slums because they are unable to make a living. Some grow the narcotic "chat" alongside coffee to make ends meet. The filmmakers do not address the role of the Ethiopian government in addressing the plight of the farmers. More recently, Human Rights Watch reports that indigenous farmers in the South are forced off their traditional lands by neocolonial and neoliberal policies in the biggest land grab since colonization. Western, Indian, and Saudi corporations are buying up massive tracts of land or securing cheap leases for corporate food production. In Ethiopia, the government is facilitating this process and hundreds of thousands of indigenous peoples have been displaced, there are reported human rights abuses, and in the process their traditions have been destroyed. In addition 70,000 coffee farmers have been forced off their lands (Mauduy & Pelleray, 2014; Westhead, 2014).

Another negative impact of coffee and other large-scale commodity production for the Western market is the environmental degradation of the land. The development of new high-yield varieties has pressured producers to turn to large-scale mono-cropping, which creates soil erosion, reliance on fertilizers, and chemical pesticides. This has negative effects on the land, wildlife, and the health of the local people (Fridell, 2012).

Multinationals spend hundreds of millions of dollars on advertising and branding to get us drinking coffee. Tim Hortons' brand is that of a caring and community-based Canadian corporation and Starbucks promotes its brand as hip and progressive. These brands have been embraced by North Americans and continue to expand. However, what is hidden from consumers is the exploitation of coffee farmers in the drive for increased profits by these Western corporations. As Western consumers we have no idea about the lives of the coffee farmers; to us coffee, like most products we consume, is an abstract commodity on the shelf in the supermarket. It is independent of exploitative economic, historical, political and social relationships (O'Brien & Szeman, 2010). Yet Tim Hortons and Starbucks are nothing like the advertising mythologies they create. Both are large multinationals, Tim Hortons has more coffee shops than McDonald's and has been owned by the U.S. corporation Wendy's since 1995. Not only does Tim Hortons and Starbucks exploit low-paid coffee workers and small farmers they also exploit low-paid non-unionized North American workers. These corporations utilize a range of methods to keep wages low. They use primarily piecework and seasonal employment, rely heavily on female workers who are vulnerable to exploitation because of patriarchal norms, and receive significantly less pay than men. Coffee servers and workers rely on unpredictable and unstable employment. In addition to non–full-time status and low pay, they get limited health and pension benefits. Most workers are from easily exploited groups such as older workers, youth, single mothers, Aboriginal Canadians, recent immigrants, and persons with disabilities. Because of poor conditions, the annual turnover rate for the coffee sector is 200 percent, among the highest in any sector of the economy. Neoliberal ideologies have allowed coffee corporations to lobby governments to lower or freeze real minimum wages, employment and welfare benefits, corporate taxes, and aggressively fight unionizations and worker's rights to collective bargaining (Fridell, 2012).

Hegemony

Domination by Western liberal democracies is subtle with less obvious benefits for dominant groups because of hegemonic practices and discourses. Dominance is secured through interpersonal, cultural/ideological, and institutional forces working together to produce systemic impacts that legitimize power and privilege. Dominant groups control institutions for their own advantage and can establish policies and procedures that can provide, deny, or limit opportunities and access to resources and power. Institutions such as media organizations, schools, entertainment, and advertisers normalize the exploitative relationships of neoliberal capitalism—we take inequality for granted and fail to acknowledge the historical conflicts and ideological contexts. We see inequality as the result of a natural global hierarchy of the "superior" West and the "backward" South. As we sip our (non-fair trade) morning coffee or flock to Tim Hortons and Starbucks, most are unaware that we are a link in the chain of corporate monopoly structures implicated in creating conditions of inequality, poverty, and displacement. Many of us want to "help" with poverty in Africa and elsewhere but have no understanding of the scope of these problems.

A shortcoming of the triangle model is that it focuses primarily on deconstructing the social problem to help us to understand the invisibility of power as expressed in ideologies and structures of power; it does not focus on agency, what individuals and groups do to challenge oppression and exploitation. One positive aspect of the video *Black Gold* is that rather than present a picture of victimization, it looks at ways that Ethiopians are challenging neocolonial and corporate control. Tadesse Meskele, general manager of the Oromia Coffee Farmers Co-operative Union attempts to bypass the multinationals by selling fair trade coffee. The fair trade movement comprises organizations committed to an economic relationship that sustains producers rather than just the profits of multinational buyers, roasters, and distributers. Growers can get up to twice the market price. It also encourages growing methods that are environmentally sustainable such as small-scale farms under forest canopies (Swift, Davies, Clarke, & Czerny, 2003).

In recent years, the growth of fair trade is being driven by multinational corporations and international institutions that use token fair trade to mask their free trade neoliberal agenda. For example, a corporation such as Starbucks gains positive publicity for selling fair trade coffee. However, only 6 percent of its coffee beans are fair trade and the other 94 percent is supplied under the old exploitative conditions. When the Ethiopian government tried to trademark its renowned coffee beans to control their brands, Starbucks was instrumental in killing this initiative. Activist organization Oxfam was able to mobilize 96,000 people to contact Starbucks through emails, faxes, phone calls, postcards, and in-store visits. Fearing damage to its brand image it was forced to sign an agreement with the Ethiopian government who then successfully registered the trademark in Europe, North America, and Japan. Ethiopia depends on coffee for two-thirds of its export earnings; an Ethiopian coffee worker earns 50 cents a day producing beans that are processed and distributed by multinationals for upwards of $26 per kilogram in the West (Fridell, 2012).

Post Viewing Reactions

After viewing documentaries that challenge dominant perspectives of social problems individuals demonstrate a range of reactions. The most common reaction is disbelief or **cognitive dissonance**. Viewers are shocked that unfair, exploitative, manipulative, and undemocratic practices actually occur in the global trading system. Most individuals are shocked to learn of the arbitrary and unfair policies of

cognitive dissonance
The resulting tension one experiences when holding on to two conflicting beliefs or struggling with new information in light of old (Gorski, 2014).

global financial institutions in enacting policies that benefit Western corporations through exploitation and displacement of indigenous peoples. In some cases, individuals deflect the issue, or deny responsibility of the powerful Western nations. Some suggest this is the way the world has always been, and the West is able to do this because of superior technology. Others focus on the growing of "chat," the narcotic used to supplement the income from coffee, and argue that they are engaging in illegal practices and could do something else. These reactions are understandable as individuals process this information. These structures of inequality we participate in and derive privilege from are invisible to us as Western consumers. We purchase commodities, and don't have to think about how and where they were produced, an example of what Goodman (2001) calls "willful innocence."

THE SELF-REFLECTIVE SOCIAL ACTION MODEL

> When we drop fear, we can draw nearer to people, we can draw nearer to the earth, we can draw nearer to all the heavenly creatures that surround us.
>
> —bell hooks

The self-reflective social action model (Figure 4.3) uses similar methodologies as the triangle model but has two notable differences. First, it begins with the individual as a subject of analysis. Its fundamental premise is that we cannot engage in social analysis unless we begin with self-analysis. We are encouraged to examine privileges, biases, stereotypes, and other perceptual filters (emotions and triggers) that hinder our quest for some measure of objectivity. The second notable difference from the triangle model (refer to Figure 4.2) is that this self-reflective model focuses on agency. It asks: what can individuals and groups do to challenge injustice, inequality, and environmental destruction?

The self-reflective social action model suggests that our understanding of social problems and social change is an ongoing process. This particular framework requires continual reevaluation of our understanding of social problems, lifelong learning about how we are implicated in perpetuating social problems, and ongoing assessment of the effectiveness of strategies for engaging in social change. As indicated in Figure 4.3, this model uses similar analytical tools as the triangle model; it emphasises the importance of examining the historical context from which social problems arise, understanding how dominant political ideologies (e.g., neoliberalism and neoconservatism) and social ideologies (e.g., white supremacy, capitalism, patriarchy, heterosexism) function through discourses to enlist our consent. In addition, we are encouraged to explore how ideologies and hegemonic discourses circulate within social, economic, political institutional

Self-Reflective Social Action Model

Dominant Ideologies/Discourses
- What are the historical and contemporary (political/social/economic) contexts of the problem?
- How are they interconnected with other problems?
- Who are the key players?
- What are the ideological positions?
- Can you recognize dominant discourses/norms/assumptions?
- How is stereotyping, scapegoating, distortions, omissions, language used?
- Who perpetuates these ideologies and discourses? Who is targeted?
- Personal/societal/environmental impacts?

Structures of Power
- How do power/privilege circulate in economic, social/political/global institutions?
- Which groups control key institutions?
- How do various institutions and structures utilize and legitimize dominant ideologies?
- Whose interest is served by current arrangements of power?
- Who benefits/who is disenfranchised from policies and legislation?
- How is wealth distributed? How are resources allocated?
- What are the human and ecological costs?

HEGEMONIC   DISCOURSES

On Going Self-Analysis
- What is your view of the issue?
- What sources inform your view?
- Relationship to the issue: insider/outsider?
- Awareness of your ideologies/worldview/biases/stereotypes?
- Privileged/marginalized social identities?
- Awareness of psychological triggers
- Denial of problem/cognitive dissonance?
- On going self-reflection/analysis....

Social Problems
- What is the problem: What claims are made about justice, freedom, equity, oppression?
- Who sees it as a problem/who does not?
- Who are in positions of power to define the problem?
- Who are the experts called upon to speak about the problem?
- Whose voices are marginalized /discredited/omitted?

Proposed Solutions
- How do different ideologies frame the issues & solutions?
- What solutions are proposed by those in power and those affected?
- Are change strategies bureaucratic: top down, legislative? Grassroots/transformative: allies, consensus, cooperative, long term?
- Does the strategy address symptoms or root causes: charity or structural/philosophical change?
- Is language/discourse used to justify bureaucratic strategies: reasonable/practical/radical?
- Are strategies contested by stakeholders, activists/government, various media?
- What are the impacts? Whose interests are served? Whose are not? What facts, events, help to answer this?

Moving Forward: Ongoing Learning
- What's changed? What have you achieved?
- Have oppressive structures changed/remerged?
- Have hegemonic discourses been reformulated?
- Have some positive outcomes been achieved?
- How have you changed? Blind spots: What do you need to keep working on/learning?
- How have you incorporated your learning into ongoing social and individual change?

Social and Individual Change
- What's your vision/goal regarding the issue?
- Has social analysis changed your understanding of how the problem can be framed differently?
- Have you sought out marginalized/omitted voices/perspectives?
- Are you considered an ally, or is your approach considered paternalistic?
- What spheres of influence does your social action involve (personal, community, institutional, political)?
- How does ideology influence your action plan/strategy?
- How are others impacted by your actions/inactions?

COUNTER   DISCOURSES

structures to establish norms and commonsense understanding about social problems and how they should be managed or eradicated.

Locating ourselves within relationships of power (intersecting privileged and minoritized identities, ideologies, and institutional structures) can be a catalyst for deep learning that can be empowering, uncomfortable, or destabilizing. When long held values and beliefs are decentred we can experience cognitive dissonance; the resulting tension one experiences when holding on to two conflicting beliefs or struggling with new information in light of old (Gorski, 2014). If we subscribe to an ideological position that denies systemic patterns of discrimination and inequality, refuse to acknowledge that social identities are socially constructed, that they confer status and minoritize individuals and groups, and buy into the hegemonic discourse that we are a post-discriminatory society, when presented with irrefutable facts that contradict our views, we may experience cognitive dissonance. Some of us will accept new insights or information and reevaluate our position, some will hold on to and defend our worldviews even though they are sexist or homophobic, or we may respond defensively as we grapple with dissonance. It is important to acknowledge these reactions as we try to make sense of the social analysis process.

Watt (2007) has identified a number of responses to cognitive dissonance. A common defensive reaction to dissonance is denial of the social problem—I have never witnessed racism or sexism, and therefore it doesn't exist. The individual's perception of their social experience becomes the universal experience of everyone else. We blame the victim, "I know many minorities who are successful; you can't get a job because you haven't looked hard enough." Or, "you are overly sensitive"; "you see racism and sexism even when it doesn't exist."

Another response to dissonance is to deflect the issue. "Well, I am white, we grew up poor, we were stigmatized and now we are successful because we worked hard. So, I am tired of hearing about how black and brown people face racism; they are not the only ones who experience discrimination." "Women are equally responsible for the sexual objectification of women. If women choose to present themselves as sexual objects in advertising what do they expect." In the first example, racial discrimination is displaced by introducing class discrimination as experienced by someone who has race privilege. Rather than looking at discrimination as intersecting, a competing hierarchy is introduced—my discrimination vs. your discrimination. This individual does not recognize that we can have privileged and minoritized identities and is deploying a privileged identity to minimize racial discrimination in hiring and in pay. It sets up a competing hierarchy of discrimination by not examining the specificity of discrimination, in this instance racism. Though oppressions intersect, they also have specific histories. In the second example the individual fails to acknowledge that the advertising industry is controlled by powerful men who are in positions to make conscious choices. They chose to represent women as sexual objects for the purpose of selling products and do so in ways that are sexist, racist, and homophobic (Whitehead, Talahite, & Moodley, 2013). Furthermore, patriarchal ideology is not only held by men; rather, both men and women can subscribe to sexist ideas or participate in the perpetuation of sexist practices and ideologies. By adopting hegemonic discourses, minoritized peoples also collude with discriminatory ideas and practices (hooks, 1994).

Another response to dissonance is false envy. "As a white person I don't have any culture, we don't even have a white history month…Your culture is so exotic, my culture is boring." These responses fail to grasp the social and political context of racialized societies and avoid a deeper exploration with superficial admiration of a minoritized culture. There is a failure to acknowledge that the dominant culture is so pervasive that it is invisible, the content of television shows, the timing of holidays, the historical perspective in textbooks, all centres on dominant Euro-North American cultural traditions (Singh, 2004).

Individuals can also rationalize social problems by invoking dominant discourses that normalize inequality. "It's human nature, throughout history we have always oppressed and exploited other groups. It's happened in the past, it will happen in the future, there is nothing we can do about it." This response ultimately leads to paralysis and resignation. This assertion is not entirely correct. As stated earlier, human societies have made choices, some have embraced inequality others have not (Bishop, 2005). Globally, societies continue to make advancements by addressing various forms of inequality. All of these gains have come about through the work of many people, usually activists who are able to get legislation and laws in place, and raise awareness to challenge discriminatory norms and practices. Change only comes through deliberate effort. As the leading African-American voice of the 19th century, Frederick Douglass said, "power concedes nothing without a demand."

Using a personal or religious principle is a response that avoids a deeper exploration of homophobia and heterosexism. In a class discussion on sexual orientation a student states, "I understand that gays and lesbians face problems but my religion is against same-sex marriage therefore it's not discrimination." Here, the individual is facing contradictory thoughts about this issue; to resolve the dissonance, they resort to a religious principle in an effort to alleviate the conflict. However, this position is still discriminatory and not different from previous discriminatory positions and norms than have been refuted (Watt, 2007).

Another response is to attack the messenger or the message (Singh, 2004). In response to the requirement to take a mandatory human rights course for the law and justice program at a community college, the predominantly white male students enrolled in the course directed their displeasure with the material by attacking the female instructors. The white female instructor was bullied and the black female instructor's car was vandalized and racial slurs were written on the classroom wall. The chair of the program responded that perhaps the course should be suspended since it was causing problems. Female, LGBTQ, and racialized faculty members are routinely rated lower on course evaluations than their counterparts; more so when teaching courses related to discrimination and inequality. The chair's reaction is an indication of the importance he places on learning such concepts and skills for the workplace. He does not acknowledge that a hostile climate has been created for the female faculty. Rather than address the racist, sexist, and criminal behaviour he blames the course for causing these problems. This is a dramatic example of denial and how it can sometimes manifest within the learning environment.

Examining our responses to information that challenges our worldviews is critical to social analysis. It is also important for personal growth. The downside of not being able to acknowledge privilege or denying discrimination and oppression is that it can dehumanize us, rob us of our humanity and the ability to develop authentic relationships with those that we consider different from ourselves (Goodman, 2001). Those with privileged identities often elevate themselves in relation to minoritized social identities and consequently feel entitled because they have an elevated sense of their place in society (Johnson, 2006).

As we develop critical consciousness through self-analysis we are better positioned to conduct social analysis and engage in social action. Chapter 8 explores the many layers of social change and social action from being a conscious consumer to becoming a community or global activist; working outside the system on structural change or trying to reform the system from within. Our level of self-awareness and how we process the information and insights from social analysis will lead to many different outcomes for social action. If we are unable to move beyond dissonance, acknowledge privilege, or hold on to discriminatory views our approach to change may have limited or even harmful impacts.

Social change strategies require a deep understanding of the issues based on self-analysis, an understanding of ideologies and dominant discourses, institutions structures, what those in power propose as change strategies, and how this differs from what is proposed by those affected by the problem. It requires that we recognize our privileges and social locations. When we skip the difficult work involved in self-analysis, change strategies can be problematic.

Some strategies focus on providing an awareness of the problem but do not move beyond awareness. Other strategies promote inclusion into an economic or political system that is fundamentally flawed because it perpetuates inequality or lack of democracy. Some strategies can be paternalistic, and focus on helping the other. "I went on a volunteer trip to Honduras. I am so touched by the gratitude of those who I helped. I can really make a difference by helping the less fortunate." This approach is based on an act of charity and the focus is on the benevolence of the volunteer. Acts of charity alleviate the symptoms of problems and are important on their own. However, ignorance of the structural causes of global inequality or other social problems helps to maintain structures of dominance and exploitation (Illich, 1968).

Problematic change strategies also include strategies that minimize the magnitude of the social problem. Many educational programs and workshops focus on understanding the "other"—how can we become culturally competent to understand minoritized people? How are "they" oppressed and how can our institutions "help" "them." These programs reduce social problems to simple facts and statistics; homogenize complex cultures by providing check lists on particular cultural nuances (Nestle, 2008; Singh, 2012). This attempt at cross-cultural exchange can reproduce stereotypes and does not address deeper systemic change. These are some of the most common pitfalls of social action/social change strategies. They give the appearance of creating social change but in reality they maintain the status quo with superficial efforts that don't change the fundamental structures of inequality and oppression.

CRITICAL THINKING QUESTIONS

1. How does understanding the concepts of ideology and discourse help us to think about our thinking processes, our assumptions, and conclusions about social problems?

2. After reviewing the dominant ideologies discussed in this chapter, can you identify which ideology has a significant influence on your worldview?

3. Can you identify any problems in your thinking? How do you respond to ideas and facts that challenge cherished beliefs? Do you expect more of others than yourself?

4. Identify a cultural or national group that you are familiar with. Next, construct a hierarchy of the social groups within it. Can you identify which groups have more power, wealth, privilege, and prestige? How are these groups characterized? Now identify the social groups with less and less power until you reach the groups with the least power. What characteristics are attributed to these groups? How do the groups with the most power keep their power? How is inequality viewed within this cultural or national group? If groups with less power come to accept their limited power and conditions of inequality, what term best describes these ruling processes? To what extent is it possible for groups with less power to increase their power or demand more equity?

REFERENCES

Abu-Laban Y., & Gabriel, C. (2002). *Selling diversity: Immigration, multiculturalism, employment equity and globalism.* Peterborough, ON: Broadview Press.

Adams, I. (2001). *Political ideology today.* Manchester, UK: Manchester University Press.

Allen, B. (2011). *Difference matters: Communicating social identity.* Long Grove, IL: Waveland Press Inc.

Anderson, S. E. (1995). *The Black Holocaust for beginners.* New York: Writers & Readers Publishing.

Arnold, R., Burke, B., James, C., Martin, D., & Thomas, B. (1991) *Educating for a change.* Doris Marshall Institute for Education and Change. Toronto: Between the Lines Publishing.

Apple, M. (1993). Constructing the "other"; Rightist reconstructions of common sense. In C. McCarthy & W. Crichlow (Eds.), *Race, identity and representation in education* (pp. 24–39). New York: Routledge.

Bannerji, H. (1995). *Thinking through: Essays on feminism, Marxism and anti-racism.* Toronto: Women's Press.

Barnes, C., & Nazim, Z. (2004). Diversity management and the legacy of Canadian multiculturalism: Moving towards a critical social framework. *International Journal of Diversity in Organisations, Communities and Nations, 4,* 1101–114.

Bishop, A. (2002). *Becoming an ally: Breaking the cycle of oppression in people.* Halifax: Fernwood Publishing.

Bishop, A. (2004). *Grassroots leaders building skills: A course in community leadership.* Halifax: Fernwood Publishing Company.

Bishop, A. (2005). *Beyond token change: Breaking the cycle of oppression in institutions.* Halifax: Fernwood Publishing.

Brock, D. (2003). *Making normal: Social regulation in Canada.* Toronto: Thomson Nelson.

Canadian Broadcasting Corporation. (2014). Retrieved from http://www.cbc.ca/strombo/news/what-a-waste-new-study-says-canadians-waste-27-billion-worth-of-food-every-year

The Chilly Collective. (Eds.). (1995). *Breaking anonymity: The chilly climate for women faculty.* Waterloo, ON: Wilfrid Laurier University Press.

Chossudovsky, M. (2003). *The globalization of poverty and the new world order.* Montreal: Global Research.

Climate Action Network. (2014). Retrieved from http://climateactionnetwork.ca/2013/11/22/canada-wins-lifetime-unachievement-fossil-award-at-warsaw-climate-talks/

Cobb, C., & Diaz, P. (2009). *Why global poverty? Think again.* New York: Robert Schalkenbach Foundation.

Conally, W. (2008). *Capitalism and Christianity, American style.* Durham, NC: Duke University Press.

Davis, W. (2009). *The wayfinders: Why ancient wisdom matters in the modern world.* Toronto: Anansi Press.

Dei, G., & Calliste, A. (2000). *Power, knowledge and anti-racism; A critical reader.* Halifax: Fernwood Publishing.

D'Souza, D. (1991, March). Illiberal education. *The Atlantic Monthly, 267*(3), 51–79.

Edward-Galabuzi, G. (2006). *Canada's economic apartheid: The social exclusion of racialized groups in the new century.* Toronto: Canadian Scholar's Press.

Fekete, J. (1995). *Moral panic.* USA: Gray Sky Books.

Francis, M., & Francis, N. (2006). *Black gold* [Motion picture]. England: Speakit Films.

Fraser, N. (1989). *Unruly practices: Power, discourse and gender in contemporary social theory.* Minneapolis, MN: University of Minnesota Press.

Fridell, G. (2012). Coffee and commodity fetishism. In D. Brock, R. Raby, & M. P. Thomas, *Power and everyday practices* (pp. 277–98). Toronto: Nelson Education.

Goodman, A. (2001). *Promoting diversity and social justice.* Toronto: Sage Publications.

Gorski, P. (2014). Retrieved from http://www.edchange.org/publications/cognitive-dissonance.pdf

Harvey, D. (2005). *A brief history of neoliberalism.* New York: Oxford University Press.

Henry, F., & Tator, C. (2002) *Discourses of domination: Racial bias in the Canadian English language press.* Toronto: University of Toronto Press.

Heywood, A. (2007). *Political ideologies, an introduction* (4th ed.). Basingstoke and New York: Palgrave MacMillan.

Hilliard, A. (1992, December/January). Why we must pluralize the curriculum. *Educational Leadership, 49*(4), 12.

hooks, b. (1994). *Teaching to transgress: Education as the practice of freedom.* New York: Routledge.

Illich, I. (1968). Retrieved from http://civicreflection.org/resources/library/browse/to-hell-with-good-intentions

James, C. (Ed.). (1996). *Perspectives on racism and the human services sector: A case for change.* Toronto: University of Toronto Press.

Johnson, A. G. (2006). *Privilege, power, and difference.* Whitby, ON: McGraw-Hill Higher Education.

Jones, A. (2014). Retrieved from http://www.gendercide.org/case_witchhunts.html

King, R., & Singh, C. (1991, July 16). Political correctness and the new right. *The Globe and Mail.*

Klein, N. (2007). *The shock doctrine: The rise of disaster capitalism.* Toronto: Knopf Canada.

Korten, D. (2001). *When corporations rule the world.* San Francisco, CA: Berrett-Koehler Publishers.

Kulchyski, P. (2014). Retrieved from http://manitobawildlands.org/pdfs/bp3cec/peguis_kulchyski_aboriginal-rights-are-not-human-rights.pdf

Kumashiro, K. (2000). Toward a theory of anti-oppressive education. *Review of Educational Research, 70*(1), 25–53.

Lee, E. (1984). *Letters to Marcia: A teacher's guide to anti-racist education.* Toronto: Cross Cultural Communications Centre.

Lee, E. (1991). Taking multicultural, anti-racist education seriously. *Rethinking Schools, 6*(1).

Lee, J., & Lutz, J. (2005). *Situating "race" and racism in space, time and theory: Critical essays for activists and scholars.* Montreal: McGill-Queen's University Press.

Lester, D. (2005). *Gruesome acts of capitalism.* Winnipeg, MB: Arveiter Ring Publishing.

Mauduy. V., & Pelleray, R. (2014). Retrieved from http://www.aljazeera.com/programmes/peopleandpower/2014/01/ethiopia-land-sale-20141289498158575.html

Mirchandani, K., & Butler, A. (2006). Beyond inclusion and equity: Contributions from transnational anti-racist feminism. In A. Konrad, J. Pringle, & A. Prasad, *Handbook of workplace diversity.* Toronto: Sage Publications. Inc.

Mullaly, B. (2010). *Challenging oppression and confronting privilege.* Toronto: Oxford University Press.

Mullaly, B. (2012). *The new structural social work: Ideology, theory, practice.* New York: Oxford University Press.

Nestel S., & Kanee, M. (2008). *Diversity and human rights in the work environment: A qualitative research study of diversity and human rights in the workplace.* Toronto: Mount Sinai Hospital.

Nestle, S. (2008). Cultural competency or anti-oppressive educational practice? Thinking through classroom practices that promote equity and diversity. (Unpublished).

O'Brien, S., & Szeman, I. (2010). *Popular culture: A user's guide.* Toronto: Nelson Education.

Parenti, M. (1995). *Against empire.* San Francisco, CA: City Lights Books.

Patterson, M. (2014). Retrieved from http://capitalresearch.org/2012/04/bono-wants-to-save-the-world-but-he-needs-your-money-to-do-it/

Paul, R. (2014). Retrieved from http://www.criticalthinking.org/pages/richard-paul-anthology/1139

Perkins, J. (2008). *The secret history of the American empire.* Toronto: Plume Books.

Piketty, T. (2014). *Capital in the twenty-first century.* Cambridge, MA: Harvard University Press.

Prasad, A. (2006). The jewel in the crown: Postcolonial theory and workplace diversity. In A. Konrad, J. Pringle, & A. Prasad, *Handbook of workplace diversity.* Toronto: Sage Publications. Inc.

Razack, S. (2002). *Race, space and the law.* Toronto: Between the Lines Publishing.

Saiz, A.V. (2014). Retrieved from http://atarazanas.sci.uma.es/docs/tesisuma/16613703.pdf

Shovel, D. (2014). Retrieved from http://www.dickshovel.com/500.html

Singh, C. (2004). *Curriculum diversity framework document.* Centre for Instructional Development, Centennial College.

Singh, C. (2004). The politics of human rights organizational change and development (Unpublished M.A., Thesis). Trent University.

Singh, C. (2008). *Faculty competencies: Educating students for global citizenship.* Centre for Organizational Training and Development, Centennial College.

Singh, C. (2012). Three phases of human rights policy struggles at an urban community college. (Unpublished dissertation, Ontario Institute for Studies in Education/University of Toronto).

Smith, A. (2014). Retrieved from http://loveharder.files.wordpress.com/2009/08/andrea-smith.pdf

Smith, D. (2004). Despoiling professional autonomy: A women's perspective. In *Inside corporate U: Women in the academy speak out.* Toronto: Sumach Press

Stewart, D., & Douglas, D. (2000). Social construction of identity. (Unpublished).

Storey, J. (2009). *Cultural theory and popular culture: An introduction.* Toronto: Pearson Longman.

Swift, J., Davies, J. M. Clarke, R. G., & Czerny, M. (2003). *Getting started on social analysis in Canada.* Toronto: Between the Lines Publishing.

Thobani, S. (2007). *Exalted subjects: Studies in the making of race and nation in Canada.* Toronto: University of Toronto Press.

Walcott, R. (2003). *Black like who: Writing Black Canada.* London, ON: Insomniac Press.

Wallerstein, I. (1989). *The modern world-system.* San Diego, CA: Academic Press.

Watt, S.K. (2007, Spring). Difficult dialogues, privilege and social justice: Uses of the privileged identity exploration (PIE) model in student affairs practice. *College Student Affairs Journal, Special Issue, 26*(2), 114–26.

West, C. (1993). *Beyond Eurocentricism and multiculturalism.* Monroe, ME: Common Courage Press.

Westhead, R. (2014). Retrieved from http://www.thestar.com/news/world/2012/01/17/at_least_70000_forced_off_their_land_in_ethiopia_rights_group_reports.html

Whitehead, S., Talahite, A., & Moodley, R. (2013). *Gender and identity: Key themes and new directions.* Toronto: Oxford University Press.

Williams, J. (2004). *50 Facts that should change the world.* New York: The Disinformation Company Ltd.

Zine, J. (2001). "Negotiating equity": The dynamics of minority community engagement in constructing inclusive educational policy. *Cambridge Journal of Education, 31*(2).

Chapter 5

Media Literacy

Sabrina Malik and Jared Purdy

LEARNING OUTCOMES

LO-1 Define media literacy in the context of inequality and social justice

LO-2 Identify forms of bias in how media messages are constructed

LO-3 Identify stereotyping through media representations

LO-4 Understand how ideological power is maintained through ownership and regulation

LO-5 Critique the effects and impact of media on society and social change

INTRODUCTION

critical media literacy
The ability to analyze and evaluate how media messages influence our beliefs and behaviours. In this process, the viewer is not a passive recipient of these messages, but an active participant who is able to critique the content communicated by media.

When we hear the word *media*, a range of formats might come to mind. Media can refer to newspapers, the Internet, magazines, television shows, blogs, social media sites like Facebook and Twitter, movies, advertisements, and so on. From the time we wake up to the moment we go to sleep, we are constantly surrounded by and engaging with different forms of media. However, the pervasiveness of the media means that we tend to take it for granted. We might assume, for example, that the news stories we consume are reliable sources of information about the world in which we live. We might dismiss advertisements as forms of media that affect other people, but not us. We might also think of social media technology as simply constituting a new way of sharing information and connecting with our friends. Furthermore, we might insist that the movies and TV shows we sit down to watch function merely as entertainment—as a way to escape after a long day. But, as we will see throughout the chapter, the media as an institution does not exist outside the realm of power, ideology, and social structures. When we think about the media, we might begin by asking some critical thinking questions. How does the media feed into existing structures of power? Whose interests does the media serve? What does the media tell us about ourselves and those around us? Answering these questions means engaging in a process called **critical media literacy**.

CRITICAL MEDIA LITERACY

We use and consume media for a number of different reasons: to gain information, to be entertained, to express our thoughts and opinions, and to keep us connected to our social network. In order to illustrate just how pervasive media is, think about the last time you travelled using public transit. Chances are you read a number of advertisements posted in the subway and on billboards, perhaps used social media to chat with a friend, checked email on your mobile device, or watched your favourite television show online. Maybe you picked up a newspaper to get caught up on local and world events. Media messages can be found everywhere, and it is difficult to escape the impact these messages have on how we view ourselves and others.

WHAT IS MEDIA?

Media can be defined as "all the forms of communication that reach a huge audience, such as newspapers, radio, TV, magazines, film, and the Internet" (Collins Gage Canadian Intermediate Dictionary, 2006).*

Although we may have many reasons to consume media, the media itself has one primary goal: to persuade us to buy things. Advertisements are not just selling us products, but also lifestyles, values, norms, and ideals about how we should be. Our exposure to media is vast. We are bombarded with approximately 3000 media messages daily, about half of which are advertisements. Young people view more than 40,000 ads per year on television alone and are increasingly exposed to advertising on the Internet and in school (Kilbourne, 2010). The availability and popularity of social media such as Facebook, Twitter, and YouTube have the potential to greatly increase our exposure to advertising and market research. Advertising creates the most revenue for all media companies; therefore, most media messages are constructed to maximize profit. These messages are not just created to sell us products, but also to sell us ideas about what it means to be human and promote unattainable ideals that can make us anxious about our current lifestyle.

Critical media literacy is an approach to thinking about media and learning to read **media text**. In this context "read" means to critically analyze the direct and indirect messages, to probe further, and ask what is not being told. What is missing can have as much or even a greater effect on our understanding as what is told. Reading is active, it is not passive.

> **media text**
> Refers not just to words, but also images, sounds, video, taken as a whole message.

"Text" refers not just to words, but also images, sounds, and video taken as a whole message. Media literacy is the ability to analyze and evaluate how media

---

* *Collins Gage Canadian Intermediate Dictionary*, 1E. © 2005 Nelson Education Ltd. Reproduced by permission. www.cengage.com/permissions

messages influence our beliefs and behaviours. The literate viewer is not a passive recipient of these messages, but an active participant who is able to critique the content communicated by the media. We are encouraged to look beyond what is immediately presented to us and find the hidden messages through the analysis of form and content. In order to become critical observers and consumers of media, it is crucial to ask who created the message, why the message is being communicated and how the message is presented to us. By asking these questions, we begin to become aware of the perspectives and biases hidden in these messages, as well as their effects on us as individuals and in society more generally.

MEDIA BIAS

Media bias generally refers to a particular slant or perspective on specific issues. Bias can be seen through a national viewpoint. For example, Canadian news media tend to focus on issues that are particular and relevant to Canadians. Bias is also closely tied to media ownership and regulation as those who control the media decide what content we have access to and what is left out through the processes of selection and omission.

If we look at media in Canada and the U. S., we can see that most of the newspapers, radio stations, and television stations are owned by a few very large corporations. This is referred to as **media consolidation**. From a critical media literacy perspective, we must ask what issues arise when one company owns so many subsidiaries. For example, are most media messages constructed to gain and maintain power by maximizing profit for these few conglomerates?

media consolidation
The process of concentrating the ownership of media outlets by a small number of large corporations.

media regulation
Government control of mass media through laws for the purpose of protecting the public interest or promoting competition among media outlets.

In addition to media ownership, **media regulation** also determines what content we have access to. In Canada, broadcast and telecommunications media are controlled by the Canadian Radio-television and Telecommunications Commission (CRTC). Companies that want to operate radio and television broadcasting networks must first apply for a licence from the CRTC. The CRTC is responsible for overseeing media content in the interest of the people. In addition, companies that want to buy or take over other broadcasting companies must receive approval from the CRTC. The problem is that the CRTC is largely powerless against corporate interests that have pressured the government and influenced policy decisions to suit their own interests (McChesney, 2008). Through media consolidation, television, radio, print media, and increasingly access to the Internet have become concentrated into the control of fewer and fewer owners. Thus, cross-production and promotion of mass media products, as well as journalistic integrity come into question when we consider the excessive amount of amalgamation of today's media outlets (see Figure 5.1).

When ownership and regulations are considered together, an analysis of ideology becomes central to understanding media literacy. An ideology is a set of beliefs or

Media Consolidation

90% of American media was controlled by 50 companies in 1983:

That same 90% was owned by only 6 companies in 2011:

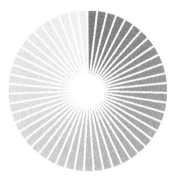

The "Big 6" are:

CBS	60 Minutes	Jeopardy	NFL.com	Showtime	Smithsonian Channel
Disney	ABC	ESPN	Marvel Studios	Miramax	Pixar
General Electric	Comcast	Focus Features	NBC		Universal Pictures
News-Corp	Fox		New York Post		Wall Street Journal
Time Warner	CNN	HBO		Time	Warner Bros.
Viacom	BET	CMT	MTV	Nick Jr.	Paramount Pictures

FIGURE 5.1 Media Consolidation

ways of thinking that shapes how we see the world. Ideologies are linked to power as they are used by those who own media to decide which representations of people and events we see and hear about. **Media representation** refers to how people, events, places, and stories are presented to us. Think of your favourite television show or film. How are various groups of people represented? Are the portrayals realistic? Are positive and negative roles distributed equally? Why is this important to think about?

MEDIA EFFECTS AND IMPACT

Most of us are resistant to the idea that media messages shape how we think and behave. We might think "That ad doesn't work on me!" without realizing that media effects are pervasive and cumulative. We don't necessarily experience the effects immediately; instead, the messages build up and become part of our worldview gradually through constant exposure and repetition. We are socialized into particular belief systems, ones that often perpetuate social inequalities. Furthermore, it takes time to reflect and critically analyze the messages we are exposed to. This is not a practice that most people engage in. In our fast-paced world, we are only able to consciously perceive about 8 percent of the roughly 3000 media messages we are bombarded with daily (Kilbourne, 2010). That means more than 90 percent of these messages are being processed below our level of awareness.

Media messages are created intentionally. The media portrays our reality; thus, much of our view of reality is based on media messages. However, the media gives us a very limited and often false representation of our world. We have the power to pick and choose which messages are truer to us or more representative of our lived reality than others. It is therefore important to access a variety of sources to include diverse perspectives and realities.

STEREOTYPING IN THE MEDIA

The media enforces and reproduces stereotypes about gender, race, class, ethnicity, religion, ability, sexuality, age, etc. These stereotypes support and perpetuate social inequalities. For example, media representations of the homeless or people living in poverty often feed into victim-blaming. These representations tend to frame homelessness or poverty as individual problems. As such, we are encouraged to believe that if we work hard enough, we can overcome any barrier regardless of our social location. Those who are struggling are often framed as leeching off social supports or not working hard enough to improve their social class. Often we see media representations of wealthy and affluent lifestyles. These do not reflect the reality of most citizens, who are working class: the vast majority of American and

Canadian citizens work in service, clerical, or production jobs (Croteau & Hoynes, 2003, p. 218). The media portrays the social world as heavily populated by the middle class, especially white middle-class professionals. These images are most obvious in advertising. Ads do not typically feature working class or poor people; they predominantly portray images of white, middle class, and affluent upper-class people (p. 218). These messages make us feel anxious about our current lifestyle while offering ways to fake our social status by purchasing luxury products and brand names. A question to consider is how these advertising campaigns feed into consumerism.

REPRESENTATION IN THE MEDIA: RACE, GENDER, AND SOCIAL CLASS

Race and the Media

The media often recreates inequalities that exist in our society. But what it chooses to reflect is very selective. An example of this is the continued portrayal of Africa as a region marked by corruption, famine and, civil war. While those issues have certainly been present throughout Africa's history, much else has gone on ranging from successful community development initiatives to private development such as the booming and prosperous movie industry in Nigeria. Similarly, seldom do we hear about Aboriginal leaders in Canada and the positive work that they have done for their communities. Usually when we hear about First Nations, the focus is on corruption on reserves, suicide, and alcoholism. In both of these examples, the media has chosen to be selective in what it shows, and as a result we don't get the whole picture; we get a picture that reinforces stereotypes that we already hold.

"Historically, the U.S. [and Canadian] media have taken 'whites' to be the norm against which all other racial groups are measured" (Croteau & Hoynes, 2003, p. 201). White culture in general is not looked at as being a culture; rather, it is normalized and becomes a standard. "Culture" becomes something that is over there, something that "others" have. Whiteness is thus assumed to be the yardstick that every other cultural group is measured against and usually found lacking. American anti-racist educator and author Tim Wise (2007) demonstrates this point when he says that we don't have a white history month, we have several, and they go by the "tricky names" of September, October, November, December, or January. Any month that is not designated as someone else's (e.g., Black History Month) is white history month. The same applies to art, music, or poetry. It is not referred to as white art, nor are the creators white artists; it is just art and they are just artists.

Historically, in the U.S., African Americans have been discriminated against in virtually every facet of their lives. When a story was told about people of African descent, it was told by a white narrator who would often portray them in the negative, stereotypical roles of nanny, servant, criminal, or buffoon living illiterate and impoverished lives.[1] In both Canada and the U.S., Aboriginals have been characterized as savages, riding bareback on horses, with bows and arrows, or as the "Indian princess." Asian, South Asian, and Middle Eastern men have often been stereotyped as sinister warlords or terrorists.

If we are to understand the topic of representation in the media, we need to consider it from three different levels: inclusion, roles, and the production process itself (Croteau & Hoynes, 2003). We can see examples of the lack of inclusion of diverse groups in all forms of mainstream media. Consider these questions: When you see news accounts of businesspeople, are they representative of the diverse demographics of our society? Which groups of people do you notice on magazine covers? When you see representations of criminals in movies, do they come from a broad spectrum of the population? This last question also considers the roles of those depicted, as in the job or activity they are portrayed doing.

While there have been gains in the frequency of appearances of racial minority groups in movies, for example, there have not been equal improvements in the types of roles they play (i.e., representation). There were not many parts for black actors in the 1930s and the few that existed were either entertainers or servants. By the time television came of age in the 1940s and '50s, their situation had changed little. In the 1960s through to the late 1990s, we gradually saw the number of black actors on television and in movies increase (Croteau & Hoynes, 2003). And yet, as bell hooks (2005) points out, there has been a continual representation of blacks in stereotypical roles where women are portrayed as promiscuous, the men are violent, and both are overly sexualized with a consistent theme of poverty and crime running through the narrative.[2]

Other marginalized groups such as Latinos, Asians, or Aboriginals, as well as women, gays and lesbians, and disabled people have been underrepresented and misrepresented in the media. As an example, Croteau and Hoynes (2005) note that Asians have had proportionally fewer and less-diverse roles in movies and daytime television, even when the location of the program has a high percentage of Asians, such as San Francisco. They also discuss the way in which Arab Amer-

---

1. Perhaps one of the best commentators on this issue in the 20th century was Marlon Riggs. Riggs was a tenured professor at UCLA Berkley and was most well-known for his analysis of the portrayal of African Americans in the motion picture and television industry in the USA. His most famous works were documentaries, one in 1987 titled *Ethnic Notions* and another in 1989 titled *Tongues United*, which garnered him international recognition. In the mid-1990s he was advocating for a much more inclusive and representative portrayal of African Americans on television and released a documentary titled *Color Adjustment*, which traces over 40 years of African American depictions in television.

2. In this context narrative is referring to the historical, social, cultural, and political contexts of the story being told. Narratives, it is argued here, are political by nature because of who gets to tell the story. This is consistent with what Herman and Chomsky have talked about in *Manufacturing Consent*, as well as in what Chimamanda Adichie discusses in "The Danger of a Single Story."

icans have been stereotyped and maligned in movies such as *The Siege* (1998) and *Rules of Engagement* (2000), which resulted in protests led by the Council on American-Islamic Relations over the racist and Islamaphobic[3] portrayals of Muslims in America.

Gender Representation and the Media

Gender in the media is no less contentious than race or social class and often intersects with both. A review of advertisements depicting women from the mid-1900s until today shows a striking display of how women have been defined, described, and characterized. This includes how they have been emotionally profiled and objectified as a body type and body parts. Women have been characterized as weak, emotionally fragile, infantile, and vulnerable.

In a 1950s print ad for a coffee company called Chase & Sanborn, there is an image of a woman lying over the lap of her husband as he is spanking her for not buying the right coffee. A Schlitz beer ad from the same era shows a woman in the kitchen with a smouldering pot on the stove. She is upset because she burned the supper. Her emotionally stable and strong husband is trying to console her when he says, "Don't worry darling, you didn't burn the beer." These advertisements reinforce the stereotype of the dutiful housewife whose job it is to serve her husband.

Advertising campaigns started to change with the rise of the women's movement in the 1960s. In many cases women were still portrayed as weak and generally incapable of functioning without a man or husband, but the images started to introduce sexual elements. From the 1970s through to today, there has been a radical change in the way women are portrayed. Advertising has become heavily sexualized. Women often appear with little to no clothing at all and are frequently reduced to nothing but a prop to showcase the item that is being sold. In other advertisements, violence and possibly rape are implied themes as was the case in several ads by U.S.-based men's clothing company, Duncan Quinn (Filipovic, 2008).

In other areas of the media, women have also made few gains. For example, in 2001, women made up only 20 percent of American news directors compared to men, while they made up 40 percent of the workforce (Croteau & Hoynes, 2003, p. 212). In the motion picture industry, women made up only 24 percent of the producers and 11 percent of the directors (Lauzen, 2001). In addition, acting roles

---

3. Islamaphobia refers to the fear or hatred of Muslims and their cultures. It can refer to general intolerance (e.g., the disapproval of Muslim women wearing the burqa or hijab, the screening of Muslim men and women at border crossings). It can also include overt racist characterizations of Muslims in popular Hollywood movies, such as *Under Siege, True Lies, Homeland,* and *Argo.* This attitude suggest that Islam is a monolithic faith, that is unchanging and universally oppressive towards women, it does not fit in with Western cultural norms, and that it is a religion of violence.

for women are more limited than those for (white) men, they have a shorter acting career with respect to age than men, they don't get paid as much as men, and are often judged by a different standard than men (Croteau & Hoynes, 2003).

Social Class and the Media

Media is a driving force of popular culture. As opposed to "high" art, popular culture is perceived as art for the masses, or "low-brow" art. This negative perception of popular culture has its historical roots in the French Revolution, when the social position of those in power was challenged by the larger population. The upper classes began relying on cultural choices as a way of distinguishing themselves from the majority. The use of cultural preferences as a form of social status was heightened by the Industrial Revolution, which began in England in the 18th century and eventually spread to the European continent and then to North America. By the 1950s, there was a belief that industrialization (and, consequently, urbanization) resulted in people losing their strong ties to community and family of origin. The critique of mass culture suggests that those who consume it (i.e., the lower classes) are distracted by sex and violence, and have lost touch with virtue and morality. They are, therefore, easier to manipulate for political and commercial gains (Ryan & Wentworth, 1999, p. 48).

MEDIA AND POWER

To disseminate messages to a mass audience requires considerable money. Money wields control, which also creates authority. Those with the means to disseminate stories, ideas, and information are in a position to influence not just which stories are told but where and how they are told. In effect, they control the narrative. Nigerian author Chimamanda Adichie (2013) reminds us that such control and one-sidedness create the conditions for what she describes as the "danger of a single story." Adichie is speaking specifically about the way in which Africa has been represented in scholarly publications, news, opinion pieces, and entertainment throughout the last 500 years. As Africa was colonized by various European powers, the Europeans were in a position to influence how many African countries and their inhabitants were described and defined. From an early point in colonization, Africans have been depicted as beasts, heathens, savages, primitive, and child-like[4]—very similar to the way in which North American indigenous peoples have been defined by their colonizers. What Adichie is also saying is that the West (Europe and America, in particular) enjoyed and continues to enjoy the ability to disseminate literature from its authors such that they have been widely read around the world. Western

---

4. See works such as Walter Rodney's *How Europe Underdeveloped Africa* and Franz Fanon's *The Wretched of the Earth.*

literature and news articles are written primarily by Americans and the British, so they get to control the narrative. This has largely not been the case for stories told about Africa and Africans.

What this analysis is also suggesting is that world powers (including their large corporations) are able to set the agenda in terms of what the public is taught in schools and what we consume on a daily basis through any number of media outlets (Herman & Chomsky, 1988). Critical media literacy encourages us to challenge dominant forms of thinking. In light of this, we ought to ask: Does our democratic system actually encourage critical thought?

NOAM CHOMSKY AND MANUFACTURING CONSENT

Noam Chomsky and Edward Herman, American media analysts, point out in *Manufacturing Consent—The Political Economy of the Mass Media* (1988) that the so-called "free press"[5] is not nearly as free as we think it is. Instead, Chomsky and Herman argue, media is constrained by economic and political decisions that can't be at cross purposes to advertisers who provide revenue to media outlets. Neither can the media be in conflict with the state because the state grants them license to operate and provides the overall legitimizing framework for profit-making businesses. Chomsky and Herman further argue that the media creates "necessary illusions" (Herman & Chomsky, 1988). These are created by broadcasting activities such as sports and other forms of entertainment along with normalizing foreign countries' problems. Normalizing issues elsewhere in the world implies that they don't need our attention. We are further distracted from trying to fix more pressing local and national problems by the illusions of happiness and tranquillity provided by the media.

The goal of manufacturing consent[6] is to create polarization in the public's mind: the world is composed of good and evil, and nothing else. Major media corporations become conduits not only for selling the capitalist dream, but also for shaping

---

5. The "free press" is simply defined as a press that is free from state control, and is an expression that usually refers to the media in Western nations (Western Europe, Australia/New Zealand, and North America). However, the term is contested because critics (i.e., Herman and Chomsky, 1988) point out the influence of the capitalist system in shaping the news and media. What this suggests is that while the "free press" may be free from the direct influence of the state in terms of censorship and control, mainstream media derives much of its revenue from advertising that is paid for by large corporations.

6. "Manufacturing consent" is an expression that was coined by American journalist and political commentator, Walter Lippman in 1922, but more widely used by Edward Herman and Noam Chomsky. It is used to explain the way in which the public gives their consent to the government and the corporations that we do business with. For example, through a process of the government's control and manipulation of certain information, they are able to manipulate messages that they then give to the media, who in turn feeds those messages to us. Similarly, Herman and Chomsky argue that the constant portrayal of a middle class, consumer lifestyle in advertising and daytime television programs influences our consent of that type of lifestyle. It actively encourages our uncritical consent to be consumers in a capitalist system.

public perception on foreign policy issues. Ideologies are built into media messages and it is our role to understand what those messages are. Without engaging in critical media literacy, we are falling in line with what the dominant media wants us to think. We need to be aware that information is subjective. We need to ask: Who owns and produces knowledge?

NOAM CHOMSKY'S AND EDWARD HERMAN'S PROPAGANDA MODEL

The propaganda model, developed by Chomsky and Herman, appeared in their 1988 book, *Manufacturing Consent*. Revisiting the model again in 1996, Herman stated that there are "crucial structural factors" that need to be underscored in an attempt to understand how the model works:

> The crucial structural factors derive from the fact that the dominant media are firmly embedded in the market system. They are profit-seeking businesses, owned by very wealthy people (or other companies); they are funded largely by advertisers who are also profit-seeking entities, and who want their ads to appear in a supportive selling environment. The media are also dependent on government and major business firms as information sources, and both efficiency and political considerations, and frequently overlapping interests, cause a certain degree of solidarity to prevail among the government, major media, and other corporate businesses. (Herman, 1996)*

Chomsky and Herman argue business entities, government, and collectives such as the Chamber of Commerce and various industry lobbies are able to exercise considerable influence over media. The authors also suggest that, "The power of the U.S. propaganda system lies in its ability to mobilize an elite consensus, to give the appearance of democratic consent, and to create enough confusion, misunderstanding, and apathy in the general population to allow elite programs to go forward" (Herman, 1996). Brewer (2011) also discusses how the general public is easily misled on any number of political issues because of the limited sources from which most people get their news. The model stresses that influence over the media is not centralized nor is it a conspiracy. Given the number of shared interests between government and corporations, there is no need for a conspiracy. "It is a model of media behavior and performance, not media effects" (Herman, 1996). Herman and Chomsky argue that there are five classes of filters that are used by corporate media to determine what is newsworthy. The filters are: corporate ownership, advertising, sourcing, negative responses to a media statement or program, and creating polar opposites.

---

*   Edward S. Herman, *The Propaganda Model Revisited*, (1996). Found at: http://www.chomsky.info/onchomsky/199607--.htm

The Five Filters

1. Corporate Ownership

Major media corporations are huge, profit-seeking entities and, as such, they have a direct bearing on editorial content. Mainstream media in the United States is composed of six large conglomerates: General Electric, News-Corp, Disney, Viacom, Time Warner, and CBS (refer to Figure 5.1). Together these six corporations control 90 percent of the media and in 2010 their combined profits were $275.9 billion (Lutz, 2012). Chomsky and Herman remind us that very wealthy and politically connected people sit on the boards of these corporations. The authors also remind us that General Electric is an enormous multinational company that is heavily involved in the production of weapons and nuclear power. Commenting on the propaganda model, David Cromwell (2002) states,

> It is difficult to conceive that press neutrality would not be compromised in these areas. But more widely, press freedom is limited by the simple fact that the owners of the media corporations are driven by free market ideology. How likely is it, then, that such owners would happily allow their own newspaper, radio or TV station to criticize systematically the 'free market' capitalism which is the source of this material wealth?*

2. Advertising

Advertising is essential to the success of any newspaper or television station because without it, they would most likely go bankrupt. In this fashion, the media is very friendly towards capitalist and corporate business interests. Chomsky and Herman point out that corporations that buy major advertising space can influence the content being printed or broadcast. These large companies have the power to withdraw their advertising dollars in the event that editorial content or programs are seen as sending contradictory messages to what is being advertised.

3. Sourcing

Sourcing is where media outlets acquire their news. To begin, the media is dependent on the government for information relating to major issues involving the government. This means that this information is highly controlled. Cromwell (2002) points out that the status of the source also has a bearing on how information is chronicled and received by the public. Senior politicians, the military, police, judges, etc., are regarded as "official" and objective or neutral; whereas strikers, picketers, and protest demonstrators are seen as heavily subjective in their statements and often portrayed as biased. Clearly that kind of framing will also determine what is considered news and, therefore, what will be printed or broadcast.

---

*   David Cromwell, (2002) *The Propaganda Model: An Overview.* Found at: http://www.chomsky.info/onchomsky/2002----.htm

4. Negative Responses to A Media Statement or Program ("Flak")

Flak refers to government or corporate counterattacks against those who criticize them. A story related to the Alberta oil sands industry is an example of how flak has been used in this way. A government flak campaign aimed at The David Suzuki Foundation (for their criticism of the oil sands project) prompted world renowned environmentalist and scientist David Suzuki to step down as the president of the board of his organization in April 2012. Suzuki stated that he was worried about his organization losing its charitable status on account of his political outspokenness (Suzuki, 2012).

5. Creating Polar Opposites

This filter has to do with how a perceived enemy is described. During the George W. Bush administration in the U.S. after 9/11, anyone who was seen to be an enemy of democracy, free market capitalism, and western values was labelled as an evil-doer or a terrorist (Brewer, 2011, p. 236). Bush defined North Korea, Iraq, and Iran as an "axis of evil" (Washington Post, 2002). Brewer (2011) goes on to explain how the White House and the major media outlets created a dichotomy of "us" versus "them" or "good" versus "evil" in the public's imagination so that the public would support the war efforts.

CONCLUSION

Throughout this chapter, we have pointed out several ways that the media works in maintaining and perpetuating current systems of power through the use of bias, representation, ownership, and propaganda. It is also important to think of how media can be subversive, in particular in the context of social change and justice. We can think of examples in our recent history such as the Arab Spring or Occupy Movement that gained momentum through social media. We will return to the idea of media as challenging systems of power when we begin our discussion of social action in Chapter 8.

CRITICAL THINKING QUESTIONS

1. Whose interests does the media serve?
2. With the influence of bias and ideology in the media, how can you separate objective information from intentional manipulation?
3. What effect does the limitation of positive news stories from developing countries have on your sense of what is happening around the world?
4. How does the media feed into existing structures of power?
5. What does the media tell us about ourselves and those around us?
6. What values, lifestyles, or points of view do you see represented or omitted in the media?

REFERENCES

Adichie, C. (2013). The danger of a single story. Retrieved from http://blog.ted
.com/2009/10/07/the_danger_of_a/single/story

Brewer, S. (2011). *Why America fights.* Toronto: Oxford University Press.

Canadian Radio-television and Telecommunications Commission. (n.d.). *Facts and figures.*
Retrieved from http://www.crtc.gc.ca/eng/home-accueil.htm.

Collins Gage Canadian Intermediate Dictionary. (2006). Definition of media. Toronto: Nelson.

Cromwell, D. (2002). The propaganda model: An overview. Retrieved from http://www
.chomsky.info/onchomsky/2002----.htm

Croteau, D., & Hoynes, W. (2003). *Media/society: Industries, images, and audiences* (3rd ed.).
Thousand Oaks, CA: Sage Publications.

Fanon, F. (1963). *The wretched of the Earth*. New York: Grove Weidenfeld.

Filipovic, J. (2008, December 11). Most disturbing ad of the year [Web log comment].
Retrieved from http://www.feministe.us/blog/archives/2008/12/11/most-disturbing-ad-of-
the-year/

Herman, E. (1996). The propaganda model revisited. *Monthly Review.* Retrieved from http://
www.chomsky.info/onchomsky/199607--.htm

Herman, E. S., & Chomsky, N. (1988). *Manufacturing consent: The political economy of mass
media.* New York: Pantheon Publishers.

hooks, b. (2005). *Culture, criticism and transformation* [PDF document]. Retrieved from
Media Education Foundation Transcript online website: http://www.mediaed.org/assets/
products/402/transcript_402.pdf

Kilbourne, J. (2010). *Killing us softly 4* [Video file]. Retrieved from http://www.thegreatplanet
.com/killing-us-softly-4-advertisings-image-of-women/

Lauzen, M. (2001). The real story on reel women. Retrieved on November 26, 2001, from www
.nywift.org/resources/status_lauzen.html

Lutz, Ashley. (2012, June 14). These corporations control 90% of the media in America. In
Business Insider. Retrieved from http://www.businessinsider.com/these-6-corporations-
control-90-of-the-media-in-america-2012-6

McChesney, R. W. (2008). *The political economy of media: Enduring issues, emerging dilemmas.*
New York: Monthly Review Press.

Rodney, W. (1972). *How Europe underdeveloped Africa.* Washington, D.C.: Howard University
Press.

Ryan, J., & Wentworth, W. M. (1999). *Media and society: The production of culture in the mass
media.* Needham Heights, MA: Allyn & Bacon.

Suzuki, D. (2012, April 13). An open letter from Dr. David Suzuki. Retrieved from http://www
.davidsuzuki.org/blogs/panther-lounge/2012/04/an-open-letter-from-david-suzuki/

University of Minnesota. (n.d.). Teaching film, television and media. University of Minnesota.
Retrieved from http://www.tc.umn.edu/~rbeach/teachingmedia/module5/2.htm

Washington Post. (2002). Text of George Bush's 2002 State of the Union Address. Retrieved
from http://www.washingtonpost.com/wp-srv/onpolitics/transcripts/sou012902.htm

Wise, T. (2007). The pathology of privilege: Racism, white denial and the costs of inequality.
Retrieved from http://www.mediaed.org/assets/products/137/transcript_137.pdf

Chapter 6

Understanding Identity

Selom Chapman-Nyaho and Alia Somani

LEARNING OUTCOMES

LO-1 Understand identity as a social construction

LO-2 Explore the relationship between identity, power, and inequality

LO-3 Identify how stereotypes are formed and perpetuated

LO-4 Consider how norms shape the categories of gender and sexuality

LO-5 Examine "race" as a social and historical process rather than a biological fact

LO-6 Understand identity as a site of contestation and struggle

Perhaps instead of thinking of identity as an already accomplished fact, which the new cultural practices then represent, we should think, instead, of identity as a 'production', which is never complete, always in process and always constituted within, not outside, representation.

—Stuart Hall, "Cultural Identity and Diaspora"*

INTRODUCTION

Identity is a complex issue. Each one of us will define ourselves differently and each one of us is unique. Indeed, we have our own fingerprints; if we are students, we might have our own student number. When we travel, we might be asked to produce our passport—or proof of our uniqueness as individuals. But our identities are also shaped by society—its structures and its institutions (the family, the school, the media, the government). This is why we might think of our identity as Stuart

---

* Stuart Hall, (1990). "Cultural identity and diaspora". In J. Rutherford (Ed.), *Identity: Community, Culture, and Difference*, Pg. 392-403, London: Lawrence and Wishart.

Hall does in the epigraph to this chapter: not as "an already accomplished fact" but instead as "a 'production', which is never complete, always in process" (1990, p. 395). In this chapter, we will address the complex ways in which our identities are shaped and defined. We will ask: What are the categories in which we identify ourselves and where do those categories come from? How do we relate to those categories? And, how do they affect the way we see ourselves and others?

Our identities are multiple and shifting. They can be understood, for example, in terms of class, sexuality, nationality, ethnicity, age, and (dis)ability. But we are not solely determined by any one of these things. Our identities are shaped by our experiences and our histories; they are responsive to social expectations and the normative structures[1] that govern our society. They are also informed by the stories we tell about ourselves or the ways in which we choose to present ourselves to others, all based on our own ideas about how we relate to or fit with the messages we receive about who we are.

THE SOCIAL CONSTRUCTION OF IDENTITY

There are two general frameworks under which different conceptualizations of identity are understood: **essentialism** and **social constructionism**. According to essentialist theories, identity is innate; it is something with which we are born. It is the idea that we have some kind of unchanging fundamental self. Identity, thus, lies outside the sphere of culture and politics (O'Brien & Szeman, 2004, p. 170). Social constructionism argues that identity is the product of social structures and culture. Whereas essentialist theories suggest that identity is wholly determined within the individual, social constructionism suggests that identities are shaped by external factors and change over time. They are, in part, determined through our interactions with social structures, social institutions, and other people. And these are all inextricably linked with power. Social constructionism understands identity not as a naturally occurring fact but rather as produced and mediated—or constructed—within the social relations of power and inequality.

In order to see how identities are socially constructed, we must understand how society functions. More specifically, we must understand the social structures that mediate our interactions. One component of this social structure is social status. **Social status** is the position a person has within a society. It is a socially defined category within a group or society that often carries certain expectations, rights, and duties (Murray, Linden, & Kendall, 2014, pp. 119–20). For example, "student" is a social status as is "doctor," "bus driver," "child-care worker," and "police officer." These defined positions carry a certain degree of recognition and have assigned to

essentialism
A perspective that assumes that aspects of our identities are innate. We are born with them and they remain fundamentally unchanged throughout our lives.

social constructionism
A perspective that argues that our identities are the product of the interplay between individual, cultural, and social structures.

social status
The position a person has within a society's hierarchy.

---

1. Normative structures refer to the socially accepted patterned behaviour in society or in any particular social institution.

them very specific duties. But social status does not just refer to occupations: "husband," "daughter," "senior citizen," and "addict" are also examples of a social status.

Social status is usually divided into two categories: achieved status and ascribed status. **Achieved status** is "a position in a hierarchy that has been achieved by virtue of how well someone performs in some role" (Krahn, Lowe, & Hughes, 2014, p. 129). Achieved statuses would include the recognition one receives for their education, occupation, accomplishments, abilities, and even hobbies. What we notice about achieved statuses is that they are usually those over which an individual has some control. We work towards achieving the recognition that comes with them. **Ascribed statuses**, on the other hand, are categories assigned to individuals, often at birth and cannot easily be changed. They are given to us involuntarily (Murray et al., 2014, p. 120). Ascribed statuses tend to include things like ethnicity, sex, and age. One thing that is similar about both achieved and ascribed statuses, however, is that society tends to assign roles to these different categories.

Roles are the social and behavioural expectations assigned to different status categories. When we meet someone on the street, we may make assumptions based on their appearance and we may group them into categories familiar to us such as race, class, gender, sexuality, age, ethnicity, nationality, and occupation. When we are asked to fill out an identification form, these same categories are typically presented to us. Moreover, we often feel pressure (consciously or unconsciously) to conform to social expectations. Thus, if we are at school where we are categorized as "students," we will often behave, dress, and carry ourselves according to certain expectations. When we go home and we are categorized as "parents" or "children," our behaviour, dress, attitude, and language might change because the expectations have changed. A role "is the dynamic aspect of a status. Whereas we occupy a status, we play a role" (Murray et al., 2014, p. 121).

Understanding social statuses and roles is important in the social construction of identity because the categories defined by a group or society, as well as the behavioural expectations that correspond to each category, structure both how we see ourselves and how we relate to other people. An important part of the social construction of identity is the notion that we learn much of our identity from our families, communities, educational institutions, and other surroundings. Over time we also learn, often without even realizing, how to selectively present ourselves to manage other people's impressions of who we are. We do not, "come into the world pre-programmed with a sense of self and the knowledge necessary to act and interact appropriately with others" (Shaffir & Pawluch, 2014, p. 51). We come to understand different social statuses and the roles, or behavioural expectations, associated with them through interactions with others. It is through this knowledge that we begin to develop a sense of self—an identity. **Socialization** is "the lifelong process of social interaction through which individuals acquire a self-identity and the physical, mental, and social skills needed for survival in society" (Murray et al.,

achieved status
A social status that is a result of an individual's work, accomplishments, and/or abilities.

ascribed status
A social status assigned to an individual from birth. It is not chosen and cannot easily be changed.

roles
The social and behavioural expectations assigned to different status categories.

socialization
The process, through interactions with others, by which we come to understand different social statuses and the roles, or behavioural expectations.

2014, p. 93). Claiming that we acquire our self-identity through socialization does not mean that we are born without any biological or psychological predispositions. Even taking into account that different children are born into certain privilege and opportunities, some children are naturally more expressive, aggressive, or sociable than others. We are born with certain dispositions, but it is through the social process of learning how to perceive and then express these dispositions that our sense of self—our individual identity—is born. Socialization begins early in our lives. Think of babies: Parents spend a considerable amount of time teaching babies how and when to sleep, eat, and communicate. Potty training is a form of socialization. How quickly individual infants acquire these skills varies, but the end result is the same. Likewise, children are socialized into behaviours such as honesty, fair play, and safety—behaviours thought to indicate good character. Children also often begin receiving religious instruction or learning cultural traditions at this time. When they are teenagers, young people are encouraged to think about possible future careers and to develop the appropriate work ethic to ensure these careers are attainable. As an example of how socialization channels biological disposition, teenagers are often also taught about the expressions of sexuality deemed appropriate in their families and communities. Socialization and re-socialization occur throughout our lives.

When talking about socialization we can see that our families are the first and perhaps most important agents of socialization. Additionally, through our families we are linked to social statuses like ethnicity, religion, culture, and social class. These statuses often determine the nature of our socialization and the particular socialization we receive often reaffirms and reproduces our social statuses. But other institutions also contribute greatly to how we learn and internalize socially appropriate behaviour, values, and identities. Schools, for instance, teach specific knowledge and skill, but they also contribute to a child's self-image and values. Some critical theorists maintain that schools also socialize, or, perhaps more accurately, discipline young people towards a set of values that contribute to future economic productivity, namely: obedience to authority, competition, and monotony. Our friends, or peer groups, also contribute to our socialization by providing a sense of belonging and reinforcing specific cultural norms (Murray et al., 2014, p. 98). An example of how influential friends can be in the formation of identities can be seen by all the attention paid in the media and society to peer pressure:

> Individuals must earn their acceptance with their peers by conforming to a given group's norms, attitudes, speech patterns and dress codes. When we conform to our peer group's expectations, we are rewarded; if we do not conform, we may be ridiculed or even expelled from the group. Conforming to the demands of peers frequently places children and adolescents at cross purposes with their parents. For example, children are frequently under pressure to obtain certain

valued material possessions (such as toys, DVDs, clothing, or athletic shoes); they then pass this pressure on to their parents through emotional pleas to purchase the desired items. (Murray et al., 2014, pp. 99–100)*

Peer pressure effectively demonstrates how socialization works to shape our identities. It is not as straightforward as being told how to think of ourselves; rather, certain values and behaviours are sanctioned or rewarded by increased social prestige, while others are either subtly or explicitly discouraged. Over time, we may internalize these ideas to the extent that they seem naturally a part of us.

One of the most powerful socializing influences on our identities is the mass media. In Canada, both children and adults spend thousands of hours every year watching television, surfing the Internet, and listening to music. While we have ever more choice in the types and amount of media we consume, as we have seen in Chapter 5, the media sends powerful messages about our values, beliefs, and position in society.

Family, schools, peer groups, and the mass media may be the most influential, but they are by no means the only socializing institutions in our lives. Religious institutions, sports teams, youth groups, support services for people suffering from addiction, correctional facilities, and even hospitals contribute to the process of teaching people how to understand themselves and act in socially acceptable ways. All of these services are designed to assist, support, and teach individuals, but they also "socialize clients to behave, think, and feel as prescribed by the institutions" (Shaffir & Pawluch, 2014, p. 68).

Socialization is fundamental to an individual's construction of their identity, but people are not simply passive recipients of the messages they receive about who they are and how they should act. We exert some agency, particularly in the way we present our identity to others. Statuses provide the socially defined categories through which we are able to identify ourselves and roles offer the behavioural expectations that correspond to each status, but each of us makes choices about how we conform to these categories and expectations. Moreover, we tend to selectively present different aspects of our identities in different settings and interactions. Most simply, much of our identity is formed through a process of noticing how others see us and reflecting on how that corresponds to how we see ourselves and how we would like to be seen. In the 1930s, sociologist George Herbert Mead theorized this process by drawing a distinction between the 'Me' and the 'I', claiming the 'Me' is oneself as others see you, whereas the 'I' is how one thinks of themselves. According to Mead, the 'I' is the source of originality, creativity, and spontaneity. The interaction between the 'Me' and the 'I' is how an individual develops both an identity and a social conscience (Craib, 1992, p. 88). Decades earlier, Charles Horton Cooley referred to the **looking-glass self** to emphasize how our identity

looking-glass self
The theory that our ideas about our identity are formed through the way we imagine we are seen by others.

---

*J. L. Murray, R. Linden & D. Kendall, *Sociology in our times* (6th ed.). Toronto: Nelson Education, 2014.

is derived from the perceptions of others. Through this looking-glass self, Cooley maintained that we imagine how our personality and appearance seem to other people and how they might judge us. This reflective process, Cooley claimed, is how we develop our identity or self-concept (Murray et al., 2014, p. 102).

Erving Goffman developed the connection between expectations about our behaviour—our roles—and our own thinking about our identities, claiming roles could be thought of as scripts from a play. He coined the term "dramaturgical approach"[2] to describe how people engage in complex performances in order to manage the impressions of others (Craib, 1992, p. 89). In his text *The Presentation of Self in Everyday Life*, Goffman details how we selectively present different aspects of our identities to different audiences. For example, Goffman wrote about **front stage** and **back stage behaviour** to distinguish the different roles we play when in public versus when we are in more intimate settings with family and friends. It is tempting to think of front stage behaviour as *fake* and the back stage as a place where a person can be his or her *real* self, but Goffman did not necessarily intend it this way. They are both our real selves. In both settings, we imagine how other people perceive us and this affects our behaviour; however, in the back stage we are less consciously anxious about how we might be judged due to the trust and familiarity we have with those closest to us.

We are each of us unique. But we are unique within a framework of socially defined categories (statuses) each of which have behavioural expectations attached (roles). Our identity, thus, is the process of negotiating these categories, recognizing how we are perceived by others, and attempting to manage others impressions of us in different social settings.

front stage behaviour

According to the dramaturgical approach, the behaviour that we exhibit when in public or around less familiar acquaintances.

back stage behaviour

According to the dramaturgical approach, the behaviour that we exhibit only when alone or around more intimate acquaintances.

STEREOTYPES, DISCRIMINATION, AND INTERSECTING IDENTITIES

While we exert some control in the construction and presentation of our identities, the behavioural expectations that are attached to different statuses often leads to stereotyping, prejudice, and discrimination. Stereotypes are unfounded and unwarranted generalizations about particular groups of people (Fleras & Elliot, 1996). They can be based on race or ethnicity, gender, religion, or other statuses.

---

2. Dramaturgical approaches employ the theatre as a metaphor for human interaction. Each individual is an actor and those around her constitute the different audiences whose opinions and impressions must be managed. For example, when a student fails a test she may say to her professor that she tried her best and did not understand the material. But then this same student may later tell her classmates that she hardly studied. The student does not want the professor to think that she is not working hard, but she also does not want her peers to think she is not smart enough to do well. The professor and her classmates constitute different audiences and in both instances she is attempting to manage their impressions. Dramaturgical analysts would suggest that impression management is so fundamental to our interactions that it is often performed unconsciously.

The notions, for example, that all Chinese people excel at math, all South Asians are good with technology, all women are emotional, or all Muslims are extremists represent unfounded simplifications about different groups. Furthermore, they are unjustified because they rely on received wisdom or casual observation rather than actual evidence. Stereotypes often persist because they are subject to confirmation bias. Once a stereotype exists, people selectively notice situations that seem to confirm or reinforce their expectations, thus perpetuating the stereotype.

Some stereotypes can seem harmless, but we must be aware of their existence particularly when they are linked to prejudice and discrimination. Stereotypes can also affect the way people view themselves in significant ways. While stereotypes are unfounded because they cannot apply to all or even most members of a particular group, they are not always negative. Stereotypes can be positive, negative, or neutral. **Prejudice**, on the other hand, "refers to negative, often unconscious, and preconceived notions about others" (Fleras & Elliot, 1996, p. 67). Prejudice differs from stereotypes insofar as they contain a moral judgment about individuals or groups. To the extent that we judge others negatively without even knowing them, we are displaying prejudiced attitudes. When people act on stereotypes and prejudiced notions, they are engaging in discrimination. **Discrimination** is "any act, whether deliberate or not, that has the intent or the effect of adversely affecting others on grounds other than merit or acquired skills" (Fleras & Elliot, 1996, p. 69). Unfortunately, all too often, people are treated differently not based on their individual qualities, but based on preconceived notions about who they are.

Different groups in all societies are subject to different levels of prejudice and discrimination. For example, people can be discriminated against based on statuses such as gender, sexuality, race, ethnicity, culture, class, caste, shade or skin colour, religion, language, (dis)ability, or body size and shape. The list is extensive. Often, people occupy two or more statuses that are subject to some degree of discrimination. We could be disadvantaged based on gender and ability, or race and sexuality, or any combination of the above. Academics and activists refer to the experience of discrimination from a number of different directions as **intersectionality**. Legal scholar, Kimberlé Crenshaw (1989), coined the term. To explain it, she used the metaphor of a collision taking place at an intersection—hence the term intersectionality: "Discrimination, like traffic through an intersection, may flow in one direction and it may flow in another. If an accident happens in an intersection, it can be caused by cars traveling from any number of directions and, sometimes, from all of them" (p. 149).

Stereotypes can lead to prejudice and discrimination, but another danger is that the groups being stereotyped can often start to believe that the negative characterizations of them are true. **Stereotype threat** refers to "poor performance in the face of negative stereotypes" (Beilock, 2010, p. 102). To examine how stereotypes can affect our behaviour and performance, social psychologists have measured how

prejudice
Preconceived negative opinions about individuals or groups.

discrimination
An act that has the intent or effect of negatively affecting others based on grounds other than merit or acquired skills.

intersectionality
The experience, or potential experience, of multiple forms of discrimination based on the intersection of different social statuses.

stereotype threat
The effect of negative stereotypes on an individual's performance or behaviour.

awareness of a stereotype can impact how individuals score on standardized tests. In one experiment, groups of Asian female college students were recruited for a math test. However, some groups were given a survey beforehand that highlighted their Asian ancestry while the others had a survey that drew attention to the fact that they were women. This test was designed to test the effect of two common stereotypes. The first, that Asians are naturally good at math and the second, that women are naturally less inclined towards the fields of science, technology, engineering, and math. The threat of stereotypes impacting performance was confirmed when the Asian women who were sensitized to their Asian identity scored higher on the test than the Asian women who were sensitized to their gender (Beilock, 2010, p. 166).

GENDER IDENTITIES AND SEXUALITY

In order to understand gender identities as socially constructed, we have to begin with a distinction between sex and gender. **Sex** is a term "used to describe the biological and anatomical differences between females and males" (Murray et al., 2014, p. 336). Thus, when we talk about the "sex" of an individual, we are usually referring to his or her biological make up as either "male" or "female," although it is important to note that these categories are not always as discrete as they may appear. Not everyone can be neatly separated into male or female sex categories. **Intersex** refers to people whose biological sex characteristics are neither typically male nor female. Instead, they exhibit elements of both. Whereas sex is linked to biology, **gender** refers to the roles of masculinity and femininity that we are expected to play based on our sex. When we say things like, "Be a man!" or, "Act like a lady!" we are talking about our gender, not about our biological sex. Gender refers to the condition of being "masculine" or being "feminine." Typically, we think of the "male" sex as "masculine," and the "female" sex as "feminine" (Lenton, 2014).

Judith Butler, a well-known feminist theorist, however, says that the assumption that "gender is the rightful property of sex, that 'masculine' belongs to 'male' and 'feminine' belongs to 'female'" (1993, p. 312) is false. What she means is that even though we are born male, for example, we may not feel or act masculine, or even though we are born female, we may not feel or act feminine. Butler's claim is that gender is not innate: we are not born with a gendered identity. Instead, gender is what she calls performative: it is constantly produced and reproduced in the way that we walk, talk, and act out pre-existing gendered roles. Gender roles are the "attitudes, behaviour, and activities that are socially defined as appropriate for each sex and are learned through the socialization process" (Murray et al., 2014, p. 306). Most of the time, most of us conform to these expectations. For Butler, this enactment of roles dramatizes the fact that *gender is a performance* and gendered identi-

sex
A term used to describe the biological and anatomical differences between females and males.

intersex
A term used to refer to people whose biological sex characteristics do not fit into the typical definitions of male or female.

gender
The roles of masculinity and femininity that we feel or are expected to play—to perform—based on our sex.

ties of "masculinity" and "femininity" comprise a set of attitudes, behaviours, and mannerisms that we learn, acquire, and even enact.

Thus, we can say that gender, unlike sex, is a social construct and that we are socialized into gender roles. From a very young age, boys are *taught* to be boys: they are encouraged to wear blue and play with trucks; while girls are *taught* to be girls: they are encouraged to wear pink and to play with dolls. Whereas girls are encouraged to be "cute" and "sweet," boys are encouraged to be "aggressive" and "strong." It is through our interactions with people (our parents, our friends, our teachers) and institutions like the media that we learn to *do* our gender correctly. Those who fail to conform to appropriate gendered roles are often ridiculed and treated as outsiders. When a little girl acts like a boy, for example, we call her a "tom boy" and assume that she will eventually adopt what we believe is the "appropriate" gendered identity. Similarly, men who are effeminate (rather than masculine) are often punished for doing their gender wrong. The expectations and norms that govern gendered identities can be oppressive for those who fail to conform to them (Butler, 2011).

When we talk about gender identities, we need to talk about power. Masculinity is stereotypically associated with the qualities of being strong, rational, independent, intelligent, and aggressive; while femininity, which is framed as its opposite, is associated with the qualities of being nurturing, emotional, dependent, kind, weak, and submissive. Of course, we must remember that these qualities are not true; they are constructed. To assume that men are naturally more intelligent than women, we might all agree, is misguided. But these stereotypes circulate in our society and, more importantly, they have significant consequences: they feed into **gender inequality.** In the workplace, therefore, we might find that men occupy more positions of power than women, in part because of the assumption that they are more capable than their female counterparts. In jobs that involve caring for children or the elderly, on the other hand, we will find more female employees. The assumption here is that women are naturally more nurturing than men. The qualities associated with masculinity tend to be understood as more valuable and more important than the qualities associated with femininity. For instance, being rational is more valuable than being emotional, being independent is more valuable than being dependent. Thus, gendered identities and the assumptions we make about them can legitimize what we call patriarchy, "a hierarchical system of social organization in which cultural, political, and economic structures are controlled by men" (Murray et al., 2014, p. 308).

Sexual preference and orientation, which together constitute our **sexuality**, are important aspects of our identities. Our sexual behaviour and expressions are guided by a set of unwritten rules that tell us to whom we should be attracted and how we can express this attraction. Thus, just as we have expectations about appropriate gender roles, we also have expectations about appropriate sexual roles.

gender inequality
Unequal perceptions, treatment, and status of groups based on their gender category.

sexuality
An individual's sexual preferences and orientation.

For example, men are expected to be sexually aggressive, women passive. Men are also granted a wider degree of latitude in terms of desire and number of partners. Whereas women, if they are proper, are expected to pursue love over physicality and to avoid the appearance of promiscuity (Lenton, 2014, p. 79). Women who assert themselves sexually are often derogatorily labelled "sluts" but no such label exists for men. These socially acceptable sexual roles not only work to limit and police appropriate sexual behaviours, they also feed into assumptions that women and men must be naturally attracted to the opposite sex. All too often, we assume that sexuality is synonymous with heterosexuality. Thus, heterosexuality has tended to be framed as normal and natural while homosexuality has been perceived as its opposite: as deviant and unnatural. In fact, until 1974 in North America, homosexuality "was considered a serious psychological disorder" (Lenton, 2014, p. 79). Just as people who fail to perform the socially acceptable gender roles are censured, so too are individuals who fail to conform to the socially acceptable sexual roles. Gays and lesbians are frequently the target of slurs, discrimination, and violence. Sexual and gender identities operate within systems of power that feed into the marginalization and stigmatization of certain groups. But it is important to recognize that many have challenged these rigid systems of categorization and conformity. Feminist groups and queer activists have drawn attention to the ways in which gender and sexual roles can be oppressive and in many cases have actively sought to subvert these roles. As we will see in the next section, the construction of identities under relations of unequal power has always prompted struggle and contestation.

RACIAL AND ETHNIC IDENTITIES

What is **race**? If we asked a group of people to define and list the various races, we would get many different answers. And yet race is often taken for granted. It seems obvious to us that there are different races and people can be divided into them based on their physical characteristics. This is not so easily the case. While there are genetic and biological differences between individuals and these differences tend to cluster around geographical locations, the concept of race is a social construct, something that anthropologists have called a set of "pseudological rationalizations based on a confusion of emotions, prejudiced judgements, and disordered values" (Montagu, 1965, p. 6). Rather than a legitimate biological category, race is a social status that has a significant effect on how we see ourselves, how we see others, and how we perceive others as seeing us. Race is a classification of humans "based on identified or perceived characteristics such as colour of skin and informed by historical and geographical context" and while these classifications may be socially rather than biologically defined, it has still become an important "basis upon which groups are formed, agency is attained, social roles are assigned, and status is conferred" (James, 2010, pp. 50–51, 285). Race is an important aspect of our identities.

race
The "socially constructed classification of human beings based on identified or perceived characteristics such as colour of skin and informed by historical and geographical context; it is not a biological classification. It is often the basis upon which groups are formed, agency is attained, social roles are assigned and status is conferred" (James, 2010, p. 285).

Even to say we do not care about a person's race demonstrates a recognition that many people do care. We must still negotiate these categories as part of our identities.

Race represents one of the social statuses, or categories, that is socially constructed but that has a significant influence on our identities. Political and social factors, more than biology, provided the categories that were adopted for the designation of different races of people and this was done historically largely to facilitate the domination and subjugation of others. In this way, race is linked to the exercise of power. Because of this, many contemporary writers on race view racial categories not as things but as sets of social relations. They write about race as a process (Gilroy, 1987; Omi & Winant, 1994) and ethnic identities as sets of relations (Comaroff, 1996). Race, as we think of it today, is a fairly recent concept. Race and ethnicity have their origins in relations of inequality and power and their construction involves struggle and contestation; the making of concrete ethnic identities is found in the minutiae of everyday practice; ethnic identities are powerful and influential to the extent that they often feel natural; and the conditions that constructed this identity are often not the same as those that sustain or continually renew the identification. Racial and ethnic identities are "wrought in the particularities of their ongoing historical construction. The substance of race is essentially an act of consciousness which means that it can never be concretely defined or decided" (Comaroff, 1996, p. 247).

The doctrine of racism over alleged inborn differences only really developed during the late 18th century. While Ancient Egyptians, Jews, Greeks, and Romans recognized and placed a great deal of importance on differences, they did not have a method of distinction by race. In ancient Greece and Rome distinctions between them and "the barbarians" were based primarily on cultural grounds. While there was some evidence that certain scholars did suggest that there were inherent behavioural differences between peoples, these ideas did not rise to any significant degree of popularity (Montagu, 1965). The idea of race became more popular and influential mainly when it became useful for purposes of slavery and colonization (Banton, 1977).

When scientists first started classifying humans into different races in the 18th century, they did so with the understanding that there were no clear lines between the different racial categories and that these categorizations did not denote any degree of superiority or inferiority. But this notion of biological difference became very useful as an explanation and justification for slavery and colonialism. The original racial classifications were arbitrary. There are no lines where one race would end and another would begin (Montagu, 1965, pp. 71, 91–92). This is one of the reasons why most people would have difficulty outlining all of the different races and deciding who would belong to each. It "is impossible to discuss the issue of race with any logic or consistency" (Hirschman, 2004, p. 386) unless we recognize it as a historical and social construct. While ethnocentrism was a common feature in most societies, racism and race are relatively modern developments.

Before the 15th century there were few if any theories or ideas about race. People displayed **ethnocentrism** to the extent that they distinguished between different tribes and cities and deemed themselves to be superior, but physical differences between different peoples did not seem to be considered a matter of great importance: "There was some tendency to seize upon physical difference as a badge of innate mental or temperamental difference, but there was no universal hierarchy of races in the ancient world" (Graves, 2001, p. 15). In ancient Greece, for example, while non-Greeks were considered savage, adopting Greek culture was sufficient to shed barbarian status. There was little evidence for the idea that ancient Hebrews, Greeks, or Romans believed in the innate superiority of any race, and Roman and Greek slavery had no relationship to race (Graves, 2001, pp. 14–20). In fact, Ancient Greek and Roman writings portrayed a particularly positive view of Africans: "a respect for their way of life and admiration for their military and political roles in the Mediterranean world" (Hirschman, 2004, p. 390).

Interestingly, views on civility and barbarity often corresponded to whom one was in opposition to at the time. Julian the Apostate (Roman Emperor Constantine's successor in 400 CE) described Africans as, "civilized, intelligent, and mild-mannered, whereas he saw Aryans and Anglo-Saxons as warlike, barbaric and cruel" (Graves, 2001, p. 20). While humans have always worked on the idea of difference, examining classical writings on diversity indicates no overwhelming racist ideology: "The raw materials were indeed present, but the evolution of the racism that we know today would require social, cultural, and scientific developments originating in the Age of Discovery and its concomitant colonialism" (Graves, 2001, p. 22).

Both the Old and New Testaments of the Bible describe interethnic interactions that indicated very different conceptions of race and difference. Primarily, ethnic identity was not absolute. Individuals and groups often moved and changed identities by interacting with and joining other groups. Culture characteristics were understood as external and acquired, and ethnic identity was fluid. Before the rise of capitalism and private property, kinship connections and occupation were the primary indicators of a person's identity. Moreover, tracing the history of interethnic interactions reveals that different peoples were not as isolated as commonly supposed. Marriage between geographically separate groups was pursued for political and economic advantage. Alexander the Great, for instance, encouraged his soldiers to intermarry and learn other cultures and languages. Studies have shown the incredible malleability of ethnic identity in Africa. Although slavery was practised in antiquity and slaves were usually outsiders, it was usually not a permanent condition because people were often set free, reclaimed by their kin, or able to purchase their freedom. The roles of slaves varied widely and many in the Medi-

ethnocentrism
The tendency to believe one's own culture is superior and evaluate all others in comparison.

terranean and Muslim world rose to positions of great political power (Smedley, 1998).

The rise of Christianity and Islam placed a new emphasis on religious community as the source of identity. But Christians, Muslims, and Jews still lived together in relative harmony with frequent intermarriage before the Crusades, Spanish Inquisition, and the rise of the Christian kingdoms. Writing of this history, one author claimed:

> What was absent from these different forms of human identity is what we today would perceive as classifications into 'racial' groups, that is, the organization of all peoples into a limited number of unequal or ranked categories theoretically based on differences in their biophysical traits. There are no 'racial' designations in the literature of the ancients and few references even to such human features as skin color. (Smedley, 1998, p. 693)*

With the possible exception of indigenous Americans, all of the groups that came to be characterized as separate races in the 19th and 20th centuries interacted in the ancient world. Chinese vases were found to be widely distributed in East Africa and the people of the Malagasy Republic are descendents of the mixture of African and Asian ancestries. Also, Greek sailors met East Africans long before the Christian era. Ethiopians were often mercenary soldiers in Mediterranean armies and Northern European slaves (Slavs) "were shipped as far away as Egypt, Syria, Saudi Arabia, and the Muslim capital at Baghdad" (Smedley, 1998).

Race as the social concept we recognize today did not begin to appear until well into the 18th century. In the 18th century, the English began having wider experiences with varied populations, and developed attitudes and beliefs that had not appeared before. These beliefs represented a new kind of understanding and interpretation of human differences. In a break from historical patterns of conquest, English settlers in North America did not assimilate the people, instead keeping them distanced. But the concept of race as we understand it today was largely due to attempts by the English to claim and control Irish populations and land. Most of us today would not think of the Irish as a "race," but the language of race as designating people into separate biological categories was first applied to groups like the Irish, Southern Italians, and Jews in order to justify the cruel treatment they received. The English had had a long history of enmity with the Irish and, as a result, generated an image of savagery that became a major part of the public consciousness. English descriptions of the Irish at the time referred to them as "human chimpanzees" and "squalid apes" (Shanklin, 1994, pp. 3–5). The purpose of these descriptions was to "prove that Ireland was such a degenerate place that only English domination would enable the inhabitants to become part of the human race" (Shanklin, 1994, p. 5). The contempt for Irish culture and people peaked during

---

*   Audrey Smedley, "'Race' and the construction of human identity," *American Anthropologist*, Issue 100 (3), Pg. 690-702, 1998.

the 16th and 17th centuries, corresponding to the time when the English were expanding into the New World.

The image of human differences based on savagery became so embedded in English thought that it served as the marker for the construction of English identity. It was an identity that was brought to the New World and imposed on the indigenous population as soon as they began to resist domination. Eventually, this ideology began to take hold in most of Western Europe:

> English notions of their own superiority were enhanced by their technological, material, and political successes, by their earlier successful split from the Catholic realm, by the early rise of merchant capitalism, the development of new forms of wealth, notions about individual freedom, property rights, and self-sufficiency, and by a growing sense of their own uniqueness even among other Europeans. (Smedley, 1998, p. 694)*

In this context, "race" emerged as an important social classification to reflect an expanded sense of difference and uniqueness.

Political and social factors, more than biology, provided the categories that were adopted for the domination and subjugation of non-Europeans. Different races were constructed to categorize the world's people and, at the time, were based on nothing more than observation and speculation (Gould, 1997). Notions of difference combined with theories of superiority and inferiority justified colonial expansion and exploitation. The concept of race was used to control, marginalize, and oppress large segments of the population. In times and places like the pre-Civil Rights era United States or Apartheid South Africa, racial and ethnic identifications structured almost all aspects of a person's life. But we must note that activists, artists, intellectuals, and everyday people have always challenged these designations and racist ideologies. Racial and ethnic identities that were originally imposed to legitimate oppression and to mark inferiority have been reclaimed as sources of pride, political struggle, and community. By rejecting the negative roles ascribed to race and finding inspiration in the histories of social and political struggle and resistance, many have embraced racial and ethnic categories as identities useful for organizing struggles for social justice.

CONCLUSION

Identities are fluid rather than static; they are shaped by the social, historical, and ideological forces around us. However, as individuals, we also play an important role in the process of defining ourselves and in constructing and shaping our own identities. Think about the ways in which we might define ourselves on a social networking site like Facebook. The pictures we choose to post, or the status updates we

---

*   Audrey Smedley, "'Race' and the construction of human identity," *American Anthropologist*, Issue 100 (3), Pg. 690-702, 1998.

decide to write, or the photographs we insist on untagging; these choices reflect the kind of image we want to present to our friends and our family, and perhaps even to the world. The point then is that we as individuals can act as agents in the process of defining and constructing ourselves, even as we are socialized into existing structures and categories. One way in which we might define ourselves is as global citizens. That is, we might think of ourselves not strictly as citizens of a nation but rather as citizens of the world whose duties and responsibilities extend beyond the boundaries of the nation-state. The scholar Anthony Appiah said that the "worldwide web of information—radio, television, telephones, the Internet" means that "we can affect lives everywhere" and "[e]ach person you know about and can affect is someone to whom you have responsibilities." In order to act as global citizens, therefore, we need to define ourselves and rethink our identities in new ways, ways that Appiah says "will allow us to live together as the global tribe that we have become" (2007, p. xiii).

CRITICAL THINKING QUESTIONS

1. Our identities are socially constructed. This means that we are not born with particular identities but rather that we acquire them through our interactions with society. How is this claim related to your own identity? What social structures and institutions (the school, the media, parents, and peer groups) have shaped your identity and how?

2. How might you or your friends conform to or deviate from gendered expectations and assumptions?

3. What are the expectations and assumptions attached to your social status? Do these expectations shift when your social status changes?

4. Consider an instance in which you or your friends have been victims of stereotyping. Did the stereotype affect you? How did the stereotype imposed upon you conflict with reality?

5. How do you define yourself? Do you define yourself in terms of race, class, gender, or sexuality? What aspects of your identity do you emphasize? What aspects do you conceal?

REFERENCES

Appiah, A. (2007). *Cosmopolitanism: Ethics in a world of strangers.* New York: W. W. Norton & Company.

Banton. M. (1977). *The idea of race.* London: Tavistock.

Beilock, S. (2010). *Choke: What the secrets of the brain reveal about getting it right when you have to.* New York: Free Press.

Butler, J. (1993). Imitation and gender insubordination. In H. Abelove, M. A. Barale, & D. M. Haoperin (Eds.), *Lesbian and gay studies reader* (pp. 307–20). New York: Routledge.

Butler, J. (2011). Your behavior creates your gender. Retrieved on December 31, 2013, from http://youtu.be/Bo7o2LYATDc

Comaroff, J. L. (1996). Ethnicity, nationalism and the politics of difference in an age of revolution. In E. N. Wilmsen & P. McAllister (Eds.), *The politics of difference: Ethnic premises in a world of power* (pp. 162–84). Chicago: University of Chicago Press.

Craib, I. (1992). *Modern social theory* (2nd ed.). New York: St. Martin's Press.

Crenshaw, K. (1989). Demarginalizing the intersection of race and sex: A Black feminist critique of antidiscrimination doctrine, feminist theory and antiracist politics. *University of Chicago Legal Forum,* 139–67.

Fleras, A. & Elliot, J. E. (1996). *Unequal Relations: An introduction to race, ethnic and Aboriginal dynamics in Canada* (2nd ed.). Scarborough: Prentice Hall Canada Inc.

Gilroy, P. (1987). *There ain't no black in the Union Jack.* London, Hutchinson.

Goffman, E. (1959). *The presentation of self in everyday life.* New York: Doubleday.

Gould, S. J. (1997). The geometer of race. In E. N. Gates (Ed.), *The concept of "race" in natural and social science* (pp. 1–6). New York: Garland Publishing, Inc.

Graves, J. L., Jr. (2001). *The Emperor's new clothes: Biological theories of race at the millennium.* New Brunswick, NJ: Rutger's University Press.

Hall, S. (1990). Cultural identity and diaspora. In J. Rutherford (Ed.), *Identity: Community, culture and difference* (pp. 392–403). London: Lawrence and Wishart.

Hirschman, C. (2004). The origins and demise of the concept of race. *Population and Development Review, 30*(3), 385–417.

James, C. E. (2010). *Seeing ourselves: Exploring race, ethnicity and culture* (4th ed.). Toronto: Thomson Educational Publishing.

Krahn, H., Lowe, G., & Hughes, K. (2014). *Work, industry, and Canadian society* (7th ed.). Toronto: Nelson Education Limited.

Lenton, R. L. (2014). Gender and sexuality. In R. J. Brym (Ed.), *New society* (7th ed., pp. 74–99). Toronto: Nelson Education.

O'Brien, S., & Szeman, I. (2004). *Popular culture: A user's guide.* Toronto: Nelson.

Omi, M., & Winant, H. (1994). *Racial formation in the United States: From the 1960s to the 1990s* (2nd ed.). New York: Routledge.

Montagu, A. (1965). *The idea of race.* Lincoln: University of Nebraska Press.

Murray, J. L., Linden, R., & Kendall, D. (2014). *Sociology in our times* (6th ed.). Toronto: Nelson Education.

Shaffir, W., & Pawluch, D. (2014). Socialization. In R. J. Brym (Ed.), *New society* (7th ed., pp. 50–72). Toronto: Nelson Education.

Shanklin, E. (1994). *Anthropology and race.* Belmont, CA: Wadsworth Publishing Company.

Smedley, A. (1998). "Race" and the construction of human identity. *American Anthropologist, 100*(3), 690–702.

Winant, H. (2004). *The new politics of race: Globalism, difference, justice.* Minneapolis: University of Minnesota Press.

Chapter 7

Tracing History: Equality, Equity, and Inequality in Canada

Agnes Gajewski and Jared Purdy

LEARNING OUTCOMES

LO-1 Discuss what constitutes an inequality

LO-2 Identify key terms and concepts in the study of inequality

LO-3 Examine four factors that contribute to inequality

LO-4 Analyze Canada's role in the preservation of inequalities

LO-5 Discuss the impact of European colonization on Canada's indigenous population

LO-6 Examine contemporary land claims issues as they pertain to self-government

LO-7 Understand the importance of sovereignty and self-government for First Nations

LO-8 Explore the importance of the various conventions, treaties, and constitutions between First Nations and the federal government

inclusion

The act of establishing an environment that fosters diversity where all members of that society are believed to be equally valued contributors and participants (Anzovino & Boutilier, 2015).

multiculturalism

The "practice of creating harmonious relations between different cultural groups as an ideology and policy to promote cultural diversity" (Anzovino & Boutilier, 2015, p. 3).

INTRODUCTION

Canada perceives itself as being an egalitarian society where all individuals enjoy equal opportunities, rights, and freedoms. We like to believe that inequality, prejudice, and discrimination are issues that have ceased to exist in our country. Structures to address fairness, such as policies, acts, statements, and departments established by government, institutions, organizations, and the media, provide us with the perception that we support and practise **inclusion**—without differentiation on the basis of race, class, gender, ability, sexuality, ethnicity, or citizenship. Internationally, Canada has long been recognized and respected for its progressive principles on matters of global citizenship and human rights (Fleras, 2012). We have played an influential role in the advancement of rights and opportunities for individuals belonging to minoritized groups globally, specifically those of women, children, visible minorities, and individuals with disabilities. We have done so by passing laws and regulations that promote inclusion and equity such as the Charter of Rights and Freedoms (1982), the **Multiculturalism** Act (1988), the Employment

Equity Act (1995), and the Ontarians with Disabilities Act (2001). The Canadian identity is commonly presented as one of diversity, accommodation of difference, and tolerance.

Despite all of this, one must question Canada's commitment to inclusion and diversity. Is tolerance sufficient? Who establishes these laws and policies and whose interests do they serve? Do we support the concepts of inclusion and diversity, or are these merely catch phrases and "**tokenistic** gestures" that serve political, economic, and social aims intended to benefit those in positions of power?

Consider the Multiculturalism Act, passed in 1988 with the intent to promote respect and uphold the differences in ethnicities, cultures, religions, languages, and heritages of Canada's citizens. The Act was originally established as a policy in 1971 by Canada's Prime Minister Pierre Elliott Trudeau, who maintained that all Canadian citizens are equal and should be treated fairly. It is arguable that the Multiculturalism Act was not necessarily introduced in the interests of minoritized groups. While appearing idealistic, the multiculturalism policy served alternative aims: to attract newcomers to Canada to increase the population and therefore develop the economy, and to improve trade relations with developing countries. Previously, Canada's immigration policy explicitly favoured immigrants from Western Europe. However, economic changes that reduced emigration from Europe forced Canada to seek other sources of immigrants to meet its labour demands. Consequently, Canada was forced to improve its relations with, and open its doors to, other nations. As Canada transitioned from a colonial state, the multiculturalism policy presented an impression of acceptance and celebration of diversity to the world (Thobani, 2007). Be that as it may, this Act was created by those in a position of power and privilege and the concept of Canada as a multicultural society was controlled and defined in European patriarchal terms.

Almost three decades have passed since the inception of the Multiculturalism Act and Canada has grown as a nation. According to the 2011 National Household Survey, over 200 ethnic origins make up Canada's population with 1 in 5 Canadians identifying themselves as a visible minority (Statistics Canada, 2011). Canada continues to distinguish itself as a nation that values diversity and promotes equal rights and freedoms for all its citizens.

Has the increase in Canada's ethno-cultural diversity transformed the dominant ideology established by the English and French who first colonized the land? How has the Multiculturalism Act fared in its promise of justice and fairness for all Canadians since 1988?

While much has changed, numerous inequalities persist in Canada. **Social inequality** "can be defined as any difference in the treatment of people on the basis of class, gender, age, ability, race, ethnicity, or citizenship. This treatment generally involves restricting people's full participation in society and limiting resources and opportunities, hence, affecting overall quality of life" (McPherson, 2013,

tokenism
The "practice of including one or a small number of members of a minority group to create the appearance of representation, inclusion, and non-discrimination, without ever giving these members access to power" (Anzovino & Boutilier, 2015, p. 6).

social inequality
Difference in the treatment of people on the basis of class, gender, age, ability, race, ethnicity, or citizenship. Generally involves restricting people's full participation in society and limiting resources and opportunities, affecting quality of life (McPherson, 2013, p. 112).

p. 112). For instance, in 2013 the Quebec government proposed a charter banning all religious symbols from the public sector. The charter would prohibit all public employees from wearing overt religious symbols including turbans, burkas, hijabs, and kippas, and those receiving services would be required to uncover their faces (CBC, 2013). The proposal was made despite the guarantee of the Multiculturalism Act of "freedom of all members of Canadian society to preserve, enhance, and share their cultural heritage" (Minister of Justice, 2003, p. 3) and more so, the assurances in the Canadian Charter of Rights and Freedoms protecting religion as a fundamental freedom.

How does the attempt to deny individuals the right to express and practise their religious beliefs undermine Canada's perceived claims of multiculturalism?

Minoritized groups in Canada continue to be marginalized and oppressed by the very government, institutions, and organizations that claim to advance justice, fairness, and diversity (Thobani, 2007).

It is comfortable and reassuring to believe that Canada is a multicultural nation that promotes and accepts diversity. How can we question such a premise when rights, freedoms, and inclusion for all citizens are embedded in Canada's Constitution? The dominant discourse in Canadian society suggests that we have achieved equality. We all have equal access to opportunities and resources, which is ensured by our human rights codes and policies along with our social services and supports.

However, a danger exists in such a belief; it silences all discussions about inequality. It normalizes all forms of discrimination and oppression by deeming them to be non-existent and therefore not a problem. Such a premise further marginalizes minoritized groups and gives power to those in dominant positions, thus perpetuating existing inequalities (Thobani, 2007).

This chapter will encourage you to think critically and question Canada's claim that it is a diverse, multicultural, and inclusive nation in a broader discussion of theoretical concepts of equality, equity, and inequality. It will conclude with an overview of Aboriginal history in Canada in order to highlight the issues discussed in the chapter.

UNDERSTANDING EQUALITY AND EQUITY

To engage in an examination of Aboriginal history we must first consider a couple of key concepts that clarify what inequality is rooted in, and by implication the various attempts at reparation for that inequality. Through this we will be able to see who are (or were) the agents of discriminatory practices against Aboriginal peoples, what that meant for the emerging Canadian society at large, and how these issues are perpetuated today.

equality
Sameness where everyone is treated the same without consideration of individual needs, circumstances, background, or history (Fleras, 2012).

Equality denotes sameness—everyone is treated the same without consideration of individual needs, circumstances, background, or history (Fleras, 2012). However, equality does not necessarily lead to fairness. Globally, individuals do not possess equal access to opportunities and resources and do not enjoy equal power and privilege. Consequently, this means that we do not start off as equal. How can we then expect equal treatment to lead to equal outcomes—the very goal of social justice? MacKinnon (1989) states:

> You can't change the relationship between those who are equal and those who are unequal by giving them the same things…the relation between the two stays the same, and it is the relation that defines the inequality. The dominant measure is set by advantaged people…sameness and difference is not the issue of inequality. It never has been. To make this the issue conceals, among other things, the way that the dominant group becomes the measure of everything, including the measure of the disadvantaged group's entitlement to equal treatment. (Anzovino & Boutilier, 2015, p. 5)*

Given that we do not all start from the same place, treating people equally preserves inequality. This inequality is defined by power and privilege, more specifically the disproportionate distribution of power and privilege in society and the differing ability to exercise that power and privilege by members of different groups. Consequently, treating everyone the same would maintain existing barriers, which prevent those in minoritized groups from gaining equal access to opportunities and resources within society.

Alternatively, **equity** acknowledges the "need to take difference-based disadvantages into consideration" (Fleras, 2012, p. 146). Equity promotes the differential treatment of individuals based on need, circumstances, experiences, background, history, and so on, in order to balance the scale. Equity is focused on achieving equality in the outcome.

The concepts of equity and equality can be explored through a range of orientations. In this chapter we will be looking at them from the perspective of **social justice**, which consists of the following belief:

> The full and equal participation of all groups in a society that is mutually shaped to meet their needs. Social justice includes a vision of society in which the distribution of resources is equitable and all members are physically and psychologically safe and secure. We envision a society in which individuals are both self-determining (able to develop their full capacities), and interdependent (capable of interacting democratically with others). Social justice involves social actors who have a sense of their own agency as well as a sense of social responsibility towards and with others and the society as a whole. (Adams, Bell, & Griffin, 1997, p. 3)†

equity
Promotes the differential treatment of individuals based on need, taking into consideration circumstances, experiences, background, history, and so on. Equity is focused on achieving equality in the outcome.

social justice
"Full and equal participation of all groups in a society that is mutually shaped to meet their needs. Includes a vision of society in which the distribution of resources [and opportunities] is equitable" (Adams et al., 1997, p. 3).

---

* C. MacKinnon, "Equality rights: An overview of equality theories. Ottawa: National Meeting of Equality Seeking Groups." In Anzovino, T., & Boutilier, D. (2015). *Walk a mile: Experiencing and understanding diversity in Canada*. Nelson Education: Toronto, 1989.

† M. Adams, L. A. Bell & P. Griffin, *Teaching for diversity and social justice*, Routledge. New York: NY. 1997.

Applying the orientation of social justice to the concepts of equality and equity prompts us to think about issues from an outcomes-based rather than process-based perspective. While the process may involve differential treatment (equity), the end result would involve all individuals having equal opportunities and access (equality).

To better understand these concepts let us consider learning disabilities. Learning disabilities can be defined through a range of perspectives, rooted in ideology. The medical model perceives disability as biologically based. Consequently, there is a focus on the limitations or deficits presented by the "*disability*" with all responsibility for learning placed on the individual (Anzovino & Boutilier, 2015; Lilly, 1992; Zuriff, 1996). Such a perspective stigmatizes disability, perpetuates stereotypes, and encourages discrimination by establishing a hierarchy of "normal" versus "abnormal" (Adelman, 1992). In contrast to the medical model, the social-construct perspective views learning disabilities as socially constructed by the dominant social group (the "abled"). This perspective implies that individuals with learning disabilities are "differently-abled," meaning that traditional ways of teaching are not best suited for that individual. Rather, the individual requires accommodations to provide him or her with equal access to learning outcomes. Such individuals are very capable of learning but they just do it differently from the dominant population (Adelman, 1992; Anzovino & Boutilier, 2015; Jordan, 2007).

These beliefs influence the ways in which students with learning disabilities are treated and taught, and in turn the learning opportunities available to them. Teachers who maintain a medical model perspective of learning disability would not individualize their program to meet the unique needs of the student, rather they would provide each student with the same program and methods of evaluation (equality), as illustrated in the cartoon (Figure 7.1). Thus, disadvantaging the student with a learning disability (the minoritized group) and advantaging those without (the dominant group). A teacher with a social-construct perspective would change the learning environment and program to address the specific learning needs of each student (equity). Students would receive what they require individually in order to have equal learning opportunities (Anzovino & Boutilier, 2015; Stainback & Stainback, 1992).

WHAT ARE INEQUALITIES AND WHY DO THEY EXIST?

The concepts of equality and equity can further be defined through an understanding of inequality and its contributing factors. Inequality consists of the unequal distribution and access to resources, services, and opportunities (Curtis, Grabb, & Guppy, 2004; Harper, 2011). These disparities are socially structured and affect the treatment of and interactions between individuals belonging to different groups in society. Within this structure, some groups are more disadvantaged than others,

In order to win this race you will need to reach the finish line in under a minute.

FIGURE 7.1

Source: © Julia Gajewski, 2014.

with position determined by socially defined categories such as race, class, gender, and ability. Those who are more advantaged have a greater capacity to access, control, and influence the resources and opportunities they need to attain for their own betterment. Consequently, they maintain greater power and privilege, which reinforces their position within society and thus perpetuates the cycle of inequality (Curtis et al., 2004). Inequalities are maintained through four factors that function together: social stratification, power and privilege, ideology, and barriers.

Factor 1: Social Stratification

social stratification
Refers to "the hierarchal arrangement of large social groups on the basis of their control over basic resources" (Kendall, 2010, p. 214).

dominant groups
Those "characterized by a disproportionately large share of power, wealth, and social status" (Jianghe & Rosenthal, 2009).

Social stratification refers to "the hierarchal arrangement of large social groups on the basis of their control over basic resources" (Kendall, 2010, p. 214). Within this arrangement, groups are classified according to their social positioning. **Dominant groups**, defined as those "characterized by a disproportionately large share of power, wealth and social status" (Jianghe & Rosenthal, 2009), are at the top of the hierarchy, whereas minoritized groups, those "within a society that ha[ve] limited access to power, resources, and social status" (McPherson, 2013, p. 116), are at the bottom. Differences in social rank lead to disparities between the capacities of individuals to access opportunities and resources and to maintain power and control in order to make decisions and exercise choice. Social positioning is relative, established according to criteria defined by those in dominant groups, who determine what is deemed to be worthy, valuable, and superior. In essence, social stratification is the legitimizing and organization of inequality. Within the hierarchy some have more rights, freedoms, and choice than others. This hierarchy is reproduced across and between generations where those who are disproportionately disadvantaged maintain that position, as do their children, due to barriers that make it very difficult to advance (Adams et al., 1997; Bottero, 2004).

class
The "relative location of a person or group within a larger society, based on wealth, power, prestige, or other valued resources" (Kendall et al., 2007, p. 655).

Class refers to the "relative location of a person or group within a larger society, based on wealth, power, prestige, or other valued resources" (Kendall, Lothian-Murray, & Linden, 2007, p. 655). Class determines one's access to rewards, resources, and opportunities. Those, in turn, influence one's level of education, income, occupation, housing, health care, and life expectancy. In a capitalist system like Canada's, class is hierarchically stratified, with individuals belonging to upper, middle, working, working-poor, and underclass orders, as identified by Gilbert and Kahl (1998) (see Figure 7.2). Given that mobility between classes can be limited, many people become trapped. Those who belong to the upper class tend to have the most wealth and power, and the greatest access to opportunities and resources. This class is also made up of the smallest number of people, approximately 3 percent of the population in Canada (Kendall et al., 2007, p. 239). According to Macdonald (2014) of the Canadian Centre for Policy Alternatives, while the richest 20 percent of Canadian families make about 50 percent of the nation's income, they hold

FIGURE 7.2　Class Structure Based on Model by Gilbert and Kahl (1998)

Class Orders	Percentage of Canadian Population (%)	Examples of Occupations
Upper Class	3	Investors, executives
Middle Class	40	Professionals, medium-sized business owners, managers, semi-professionals, crafts people, non-retail sales
Working Class	30	Low-skilled manual labourers, clerical workers, retail sales
Working Poor	20	Unskilled labourers, lowest paid manual labourers, service workers
Underclass	7	Unemployed, part-time menial jobs, public assistance

Source: Data retrieved from Kendall, Lothian-Murray, and Linden, (2007) Pg. 239-241.

70 percent of the nation's net wealth (p. 8). The largest portion of the population in Canada, around 40 percent (Kendall et al., 2007, p. 240), make up the middle class, which can further be subdivided into upper-middle and lower-middle. The distinguishing factor of the middle class is education, with most individuals having completed a college or university program. The working class, about 30 percent of the Canadian population (Kendall et al., 2007, p. 241), is made up of labourers such as factory, office, and daycare workers. Approximately 20 percent of the Canadian population consists of the working poor (Kendall et al., 2007, p. 241). The people in this group live on an income level that falls around the poverty line.[1] Many individuals belonging to minoritized groups make up this class category, including a disproportionate number of single mothers and children (Kendall et al., 2007). Individuals belonging to the underclass are often unemployed and have low levels of education. Living below the poverty line, most rely on social assistance and services to meet their basic needs. Often, those who are discriminated against, marginalized, and oppressed fall into this class category, as they are unable to get jobs because of systemic and social barriers that exist in society.

While class is, for the most part, economically based, it has a direct and substantial influence over other social issues. Grabb (2004) presents the following argument:

> Class differences represent one manifestation of the more general structure of power that is responsible for generating the overall system of inequality in most

---

1. The poverty line, or low income cut-off, refers to "income thresholds below which a family will likely devote a larger share of its income on the necessities of food, shelter and clothing than the average family." For 2011, the low income cut-off for a family of four was $30,487 (Statistics Canada, 2013).

societies. Other crucial manifestations of inequality, such as those based on gender and race, can therefore be understood as a result of differential access to the different forms of power or domination in society. (p. 10)*

Consequently, those who are working poor or underclass have, on average, a shorter life expectancy, poorer health care, higher teenage pregnancy rates, higher obesity rates, higher infant mortality rates, lower literacy rates, and a greater likelihood to fall under the control of the criminal justice system (Fraser, 2012).

Relatedly, a number of scholars have suggested that the traditional class categories proposed by Gilbert and Kahl are outdated in the new economy (Savage, Devine, Cunningham, Taylor, Li, Hjellbrekke, Le Roux, Friedman, & Miles, 2013). Their model considers social, cultural, and economic capital and includes seven classes: elite, established middle class, technical middle class, new affluent workers, traditional working class, emergent service workers, and precarious proletariat (Savage et. al., 2013) (see Figure 7.3).

A study conducted by Hulchanski (2010) of the University of Toronto found that the city of Toronto could be divided, with regions based on social class, into three

FIGURE 7.3   Seven Social Class Model

Social Class Category	Description
Elite	Very high economic capital (especially savings), high social capital, very high highbrow cultural capital
Established middle class	High economic capital, high status of mean contacts, high highbrow and emerging cultural capital
Technical middle class	High economic capital, very high mean social contacts, but relatively few contacts reported, moderate cultural capital
New affluent workers	Moderately good economic capital, moderately poor mean score of social contacts, though high range, moderate highbrow but good emerging cultural capital
Traditional working class	Moderately poor economic capital, though with reasonable house price, few social contacts, low highbrow and emerging cultural capital
Emergent service workers	Moderately poor economic capital, though with reasonable household income, moderate social contacts, high emerging (but low highbrow) cultural capital
Proletariat	Poor economic capital, and the lowest scores on every other criterion

Source: M. Savage, F. Devine, N. Cunningham, et al., "A new model of social class? Findings from the BBC's Great British Class Survey experiment," Sociology, 47(2): 219–50, 2013.

---

*   E. Grabb, "Conceptual issues in the study of social inequality." In J. Curtis, E. Grabb & N. Guppy, *Social inequality in Canada: Patterns, problems, and policies* (4th Ed.). Pearson Publishers: Toronto, 2004, Pg. 10.

distinct cities (Figure 7.4). He noted that, since 1970, based on income, the lower class neighbourhoods increased to cover more than half the city while middle class neighbourhoods decreased by 37 percent. In city #1, "the average income earned by individuals (15 and older) in 2005 was $88,400," whereas in city #3, "the average income earned by individuals (15 and older) in 2005 was $26,900" (p. 6).

A similar trend was found by the United Way of Toronto (2004). Individuals who make up the population of the most poverty-stricken neighbourhoods include children, single parents, newcomers, and visible minorities (United Way Toronto, 2004) (see Figure 7.5, page 135). These are individuals with the least power and privilege in society and therefore face the most inequalities.

Those who belong to the working poor and underclass are often assumed to be responsible for their perceived failures. They are labelled lazy and incapable without considering the systemic, social, and ideological factors that contribute to the inequalities they experience (Kendall et al., 2007). While individuals can potentially move between class categories, the lower someone falls within the social hierarchy, the greater the barriers that must be overcome. Not least of these are the stigmas attributed to lower classes by higher classes and discrimination, both of which often prevent social mobility.[2] Kendall et al. (2007), based on the work of social analysts, suggest that it is more difficult for individuals of colour to advance within the class system, as racism continues to disadvantage them despite their level of education or income.

Factor 2: Power and Privilege

The second factor contributing to inequality consists of the related concepts of power and privilege. Power is "the capacity to command resources and thereby to control social situations." Resources can be material, human, and ideological and their control can be economic, political, and ideological (Curtis et al., 2004, p. 10). Power can also be defined as "the ability of people or groups to achieve their goals despite opposition from others" (Kendall et al., 2007, p. 237). Power promotes structures of dominance and subordination to be established among social groups (Kendall et al., 2007; McMullin, 2004, p. 29) and allows for inequalities to be reproduced in society. In a stratified system, dominant group members have the greatest degree of power, meaning they have the ability to make decisions, influence systems, and establish dominant ideologies, thus contributing to and maintaining inequalities. They can use their position of power to develop laws and policies that

---

2. Social mobility refers to "the movement of individuals or groups from one level in a stratified system to another" (Kendall et al., 2007, p. 663).

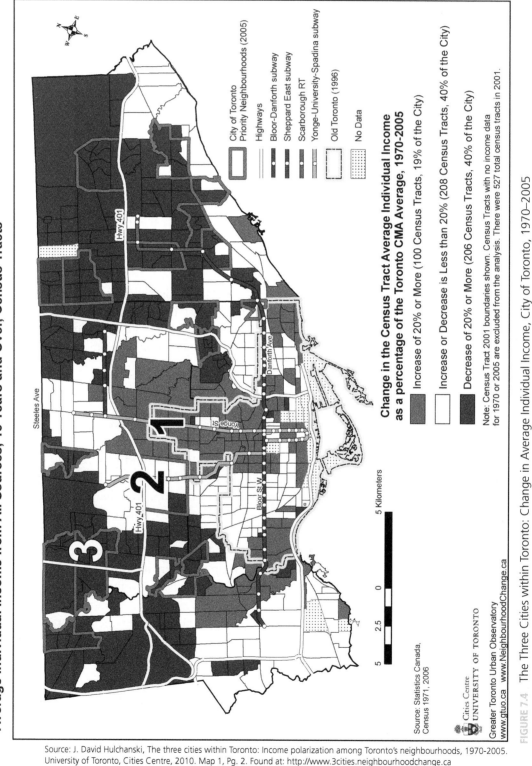

Change in Average Individual Income, City of Toronto, 1970 to 2005

Average Individual Income from All Sources, 15 Years and Over, Census Tracts

City of Toronto
Priority Neighbourhoods (2005)

Highways
Bloor-Danforth subway
Sheppard East subway
Scarborough RT
Yonge-University-Spadina subway

Old Toronto (1996)

No Data

Change in the Census Tract Average Individual Income as a percentage of the Toronto CMA Average, 1970-2005

Increase of 20% or More (100 Census Tracts, 19% of the City)

Increase or Decrease is Less than 20% (208 Census Tracts, 40% of the City)

Decrease of 20% or More (206 Census Tracts, 40% of the City)

Note: Census Tract 2001 boundaries shown. Census Tracts with no income data for 1970 or 2005 are excluded from the analysis. There were 527 total census tracts in 2001.

Source: Statistics Canada, Census 1971, 2006

Cities Centre
UNIVERSITY OF TORONTO

Greater Toronto Urban Observatory
www.gtuo.ca www.NeighbourhoodChange.ca

FIGURE 7.4 The Three Cities within Toronto: Change in Average Individual Income, City of Toronto, 1970–2005

Source: J. David Hulchanski, The three cities within Toronto: Income polarization among Toronto's neighbourhoods, 1970-2005. University of Toronto, Cities Centre, 2010. Map 1, Pg. 2. Found at: http://www.3cities.neighbourhoodchange.ca

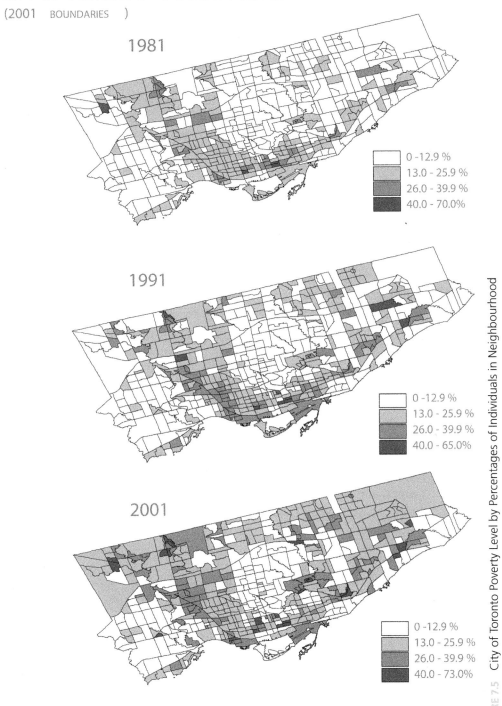

CITY OF TORONTO - POVERTY LEVEL BY NEIGHBOURHOOD
(2001   BOUNDARIES   )

1981

0 -12.9 %
13.0 - 25.9 %
26.0 - 39.9 %
40.0 - 70.0%

1991

0 -12.9 %
13.0 - 25.9 %
26.0 - 39.9 %
40.0 - 65.0%

2001

0 -12.9 %
13.0 - 25.9 %
26.0 - 39.9 %
40.0 - 73.0%

POVERTY BY POSTAL CODE

FIGURE 7.5 City of Toronto Poverty Level by Percentages of Individuals in Neighbourhood

Source: United Way Toronto and The Canadian Council on Social Development, Poverty by postal code: The geography of neighbourhood poverty city of Toronto, 1981–2001. 2004. Found at: http://www.unitedwaytoronto.com/downloads/whatWeDo/reports/PovertybyPostalCodeFinal.pdf

work in their favour and sway public views and principles to uphold their agendas (Curtis et al., 2004).

Minoritized groups within society are relatively powerless and are often vulnerable to exploitation by those in dominant groups (Kendall et al., 2007). If we consider the role of women in society, for instance, we can begin to understand the power dynamic of gender and its function. Fewer women hold high-ranking positions in the Canadian political system or in corporate and academic boardrooms. Only one third of full-time university faculty are women and only one quarter are Canadian research chairs. Sexism in academics is still prevalent and normalized through the dominant discourse (Robbins, 2012). In 2003, a human rights complaint was brought forward by eight women to expose the inequities faced by women in academia. The justice system is expected to strive for equity, to uphold human rights, and err on the side of those who have been disadvantaged and perhaps victimized. Instead, the gap was legitimized, claiming "excellence" over "equity," implying that men are simply more qualified and better suited for the positions (Robbins, 2012, pp. 1–2). Such statements, especially when made by those in power, allow—and may even encourage—the injustice and inequity to continue. Furthermore, it silences all others who face similar inequities.

Women are consistently marginalized and excluded from positions of power within a patriarchal society, where gatekeepers, primarily men, prevent their access. On the other hand, women are overrepresented and contribute significantly more to domestic labour and child/elderly care than men (Anzovino & Boutilier, 2015). This translates to the socio-economic position of women within the social hierarchy, where a significant disparity exists between men and women both in terms of social status and wealth. In 2008, the average income of women in Canada was $30,100, about 65 percent of the average income of men, $47,000 (Statistics Canada, 2013). Similar social and economic marginalization is experienced by members belonging to other minoritized groups such as Aboriginal peoples, racialized groups, and people with disabilities. Who then is making the most important decisions for Canadians? Whose interests are those decisions serving? And how do these decisions contribute to the cycle of inequality faced by many Canadians?

Membership in dominant groups also grants individuals privilege. "Privilege is gained through unearned power that gives dominant group members economic, social, and political advantage" (Lopes & Thomas, 2006, p. 266). This privilege is not deserved or attained due to hard work or achievement. Rather, it is assumed merely because one belongs to a dominant group. Those who are privileged therefore have opportunities, resources, rights, choices, and freedoms that are categorically denied to others (McPherson, 2013). Those with privilege often fail to recognize their advantaged status, as privilege in dominant groups has become so socially normalized that it has become invisible. Privilege that is unrecognized and invisible to a group equates to the degree of power that group maintains within society (Sensoy & DiAngelo, 2012). Dominant forms of privilege are reinforced socially and ideologically, thus appearing universal. For example, nationally recognized holidays in Canada, supported by government, institutions, corporations, and media, overwhelmingly follow Christian practices and celebrations. Consequently, those in minoritized groups often internalize their position and perceive the power and privilege maintained by dominant groups as status quo and fixed (Bell, 1997). Sensoy and DiAngelo (2012) point out that our vocabulary emphasizes this perception, "while we refer to the minoritized group as *underprivileged* or *disadvantaged*, we rarely talk about the dominant group as *overprivileged* or over *advantaged*" (p. 74). Discussions of privilege are completely silenced and ignored. As a result, those in dominant groups continue to benefit from the unearned privilege they maintain while those in minoritized groups continue to be oppressed by that privilege.

Differences in power and privilege are manifested by components of our identity. As discussed in Chapter 6, identity is socially constructed. These constructs determine what is considered to be valuable, superior, and powerful within society. Groups are categorized based on components of their identity, with some socially deemed as dominant and others minoritized. However, identity is multidimensional and complex. As such, we occupy more than one social group simultaneously (intersectionality) (Sensoy & DiAngelo, 2012), meaning that we can have power and privilege and we can also lack it. A man with a disability, for example, maintains power and privilege in society due to his gender status, but lacks that power and privilege due to his ability status. All of us have power and privilege due to some component of our identity. "Most of us have one or more dominant identities. In most parts of Canada, dominant identities are White, male, English-speaking, heterosexual, able-bodied, Christian, affluent and middle class, thirty to sixty-five years of age, university educated, from central Canada" (Lopes & Thomas, 2006, p. 266). Nonetheless, we can still be oppressed.

Minoritized groups, however, experience several forms of oppression that are compounded due to different and multiple components of their identity, which are denied power and privilege (Sensoy & DiAngelo, 2012).

Factor 3: Ideology

Ideology refers to systems of ideas transmitted through the process of socialization, including through systems and structures such as the media, government, families, religion, and education. The degree to which each of these systems and structures can affect one's socialization will depend upon the degree to which they are important and prevalent in a particular historical period. Different interpretations can be drawn from a study of ideology, namely that there is more than one ideology at work in society at any given time, and there are conflicts and tensions between them. They are also historically based and socially stratified along class, gender, race, and other areas of identity.

Dominant ideologies are those that are most closely aligned with a group of people who represent a particular set of class, economic, cultural, political and sociological interests situated at the top of the hierarchy. Dominant ideologies in society originate from a group of people who were directly involved in, or whose cultural ancestors were an integral part of, the drafting of a country's constitution, its charter, laws and the systems of justice, its dominant religion, and the content of the canons in educational institutions. These are the ideologies that serve to influence how society functions and how it is structured. Ideology is used by those in dominant positions to achieve desired ends, which often benefit a select few.

Ideologies serve to shape and reinforce stereotypes and prejudices. Norms and values of a society are created and reinforced by ideologies. Ideology serves to create a routine response to how certain phenomena (social, political, etc.) and certain groups of people are viewed. Dominant ideologies consistently reinforce such ideas, ultimately affecting how we see and treat others. We adopt these views through socialization and structures. Our ideas about minoritized groups in society arise due to the limited knowledge and narrow understandings we have of individuals belonging to these groups. We consider anyone or anything that deviates from the *dominant* or *norm* as different, odd, undesirable, or wrong.

Dominant ideologies hold little merit in terms of accuracy or truth. They become dominant merely because those who hold positions of power and privilege created them. They do, however, serve to perpetuate oppression and inequality. Ideologies are internalized by both dominant groups and minoritized groups within society, and as such they establish how things should be, normalizing and privileging some ideas or ways of being over others. Dominant groups control ideology and maintain power by enforcing their ideologies onto individuals, thus elevating their own status within society (Sensoy & DiAngelo, 2012).

> In order to oppress, a group must hold institutional power in society. In this way, the group is in the position to impose their worldview on others and control the ideas (ideologies), political rules (the technical mechanisms), and social rules for communication (discourses) that we are all taught (socialized) to see as normal, natural, and required for a functioning society. This domination is historical (long-term), automatic, and normalized. (Sensoy & DiAngelo, 2012, p. 52)*

---

*   O. Sensoy & R. DiAngelo, *Is everyone really equal? An introduction to key concepts in social justice education.* Teachers College Press: New York, 2012.

Factor 4: Barriers

The final factor that contributes to inequalities is **barriers**. Barriers can be defined as "policies or practices that prevent full and equal participation in society; barriers can be physical, social, attitudinal, organizational, technological, or informational" (Anzovino & Boutilier, 2015, p. 262). For example, for a visually impaired individual, a physical barrier in education may be the small font in a text or poor lighting in the classroom, and an attitudinal barrier may be the perception that the visual impairment also implies an intellectual impairment. People belonging to minoritized groups in society frequently encounter these barriers, which affect their ability to access the opportunities, resources, and benefits of membership in society (McPherson, 2013, p. 123).

For an individual with a disability, barriers "make it difficult—sometimes impossible…to do the things most of us take for granted—things like going shopping, working, or taking public transit" (Ministry of Economic Development, Trade, and Employment, 2013).

Barriers prevent or limit people from achieving equality in society and can be *visible* physically or in policies, or *invisible* in social structures. Visible barriers include those public and private spaces that are literally inaccessible to individuals who have a physical disability. Visible barriers also include the documented policies and practices in any economic, political, or social institution. These policies and practices largely represent the interests of members of dominant groups. Those in non-dominant groups are restricted from participating in policy development and therefore are not represented in it. An example is a hiring practice that prevents or restricts access to employment opportunities for individuals with disabilities (Ministry of Economic Development, Trade, and Employment, 2013). This demonstrates a visible barrier because such policies are in place to favour abled individuals and deny access to those who do not meet the criteria (McPherson, 2013, p. 123).

Alternatively, invisible barriers may not be obviously outlined and defined in laws and policies, but are maintained within social structures. Continuing with the previous example, due to invisible barriers, individuals with disabilities encounter significant challenges when they try to access employment. Often, they are perceived as inferior, deficient, or incapable. Therefore, many are underemployed, unemployed, or seek re-training (McPherson, 2013, p. 123).

It is important to note that these barriers are not internal. It is not the disability, race, or gender that is a barrier. Rather, barriers arise from the ideologies established by those in dominant groups and the individual, institutional, and structural responses to those ideologies, dominant discourses, the lack of power and privilege possessed by minoritized groups, or one's position on the social hierarchy. These barriers are problematic because they prevent members of minoritized groups from equally participating and benefiting in society, thus producing inequalities.

barriers
These "policies or practices that prevent full and equal participation in society; barriers can be physical, social, attitudinal, organizational, technological, or informational" (Anzovino & Boutilier, 2015, p. 262).

An Examination of Aboriginal Issues in Canada

Equity and Equality in Canada: Are We There Yet?

Inequalities have historical roots. They often begin with an event that serves to marginalize, oppress, and disadvantage a particular group of people. We cannot begin to understand inequalities that exist in Canada without examining how and why they came to be. As members of Canadian society, we face various inequalities and, knowingly or unknowingly, we contribute to the oppression of others. It is our responsibility, as global citizens and critical thinkers, to disrupt social hierarchies, expose dominant ideologies, unearth power and privilege, challenge barriers, and, ultimately, to uncover the causes of inequalities.

In the next section we will examine the history of Aboriginal[3] peoples in Canada. This is in order to clarify the concepts covered in this chapter and to understand how ideologies have informed behaviours and shaped the social structures that have been historically controlled by European "settlers."[4] The history of Canada has been marked by deeply held and prevailing ideas: that Aboriginal peoples are inferior to Europeans, that they need to be saved, and that they need to be integrated into the dominant culture. This history is paramount in its contribution to social issues and inequalities faced by Aboriginal peoples today. Historical attitudes have also served to maintain other existing economic, political, and social problems.

Throughout this chapter we have been discussing the issue of equity based on race, gender, sexual orientation, and other minoritized social identities. When discussing equity-related issues, however, there has been a tendency to discuss the First Nations alongside issues of equity and multiculturalism. First Nations, however, assert that their prime issues or areas of concern are the unsettled land claims and sovereignty rights as Canada's original inhabitants, who were displaced by European colonization and imperialism. For many First Nations, if there is going to be a discussion of equity for various discriminated groups of people, then it needs to include a discussion about sovereignty.

Within the dominant discourse, Aboriginal people have been stereotyped as "savage, poor and unlettered" (Axtell, 1998, p. 26), an attitude beginning as far back as 1610 (Axtell, 1998). Other references allude to colonial administrators referring to Aboriginals as children (Dickason & Newbigging, 2010, p. 160), incapable of understanding property rights. That discourse was effectively used to justify the theft of Aboriginal land and their ultimate subjugation, first to the British Crown, and then to the Canadian government. Defining Aboriginals in that fashion became increasingly important because it was easier to (morally) justify seizing land from "savages" than from "civilized" people.

The discourse around First Nations' rights, identity, and sovereignty continues to be controlled by the government, mainstream media, sports conglomerates, and fashion houses, all presenting a Eurocentric view of indigenous people. This goes on in the context of movies (white actors playing Native people, such as in *The Lone Ranger*), negative portrayals of indigenous people's socio-economic and political issues in the news, sports teams (the Washington Redskins football team's racist logo and name), fashion houses (Luis Vuitton's appropriation of Aboriginal culture such as headdresses and beaded moccasins to sell their European clothing),

---

3. For the purposes of this chapter, the terms, Aboriginal, First Nations, and indigenous are used interchangeably.
4. The term "settler" is a contested and controversial term within discussions around the colonization of Canada, or any other country that has undergone colonization. Settler implies someone who is engaged in a passive, neutral activity, with legitimate access to land that really does not belong to them. It implies other things as well, such as being invited, performing a service in the interests of the government of the country, being peaceful, and respectful. The history of colonization in Canada however, suggests quite the opposite. What is positioned in this chapter is that "settlers" were part of the colonial government's attempt at stealing Aboriginal land, and they often did so in very violent ways. They were part of the process of disenfranchising indigenous people.

and so forth. Most of these representations are historical, where the only characteristics that seem to matter are beads and tomahawks. This misrepresentation ignores vibrant and modern Aboriginal cultures and also ignores the numerous issues facing Aboriginal people. Instead, it romanticizes and commodifies their history, culture, and contemporary issues.

Historical Context: Colonization and Imperialism

Aboriginals have endured loss of land, loss of culture, and loss of basic human rights due to colonization, war, and exploitation. European expansionism through the 1600s and 1700s was the driving force behind this, and resulted in exploitation and the disenfranchisement of indigenous peoples from virtually every facet of their lives.

While there were a number of circumstances that had a drastic effect on Aboriginal ways of life, their increased involvement in the fur trade and the repercussions of that involvement had several consequences: exposure to disease, cultural genocide, engagement in a commercial wage-based economy, legislated institutionalized inequality, and their ultimate disenfranchisement from their traditional ways of life.

Impacts of Colonization: Disease

First Nations in Canada were particularly hard hit when exposed to European diseases that they were not immune to. The increased exposure was due to the increase in trade and contact between Europeans and First Nations as time went on. For example, the establishment of fur trading posts[5] along the coast of British Columbia (and in other parts of Canada) created fertile grounds for the spread of diseases, such as smallpox, cholera, measles, tuberculosis, and venereal disease to once-resilient communities (Harris, 1998). These diseases occasionally raged in epidemic proportions and wiped out as much as half the population in a given area (Gibson, 1980, p. 236).

The British Columbia provincial government even used the decimation of First Nations' population in their calculations to determine what kinds of treaty provisions First Nations would be entitled to. For example, they figured that, if the population of a treaty signatory was small, the government could pay them less money or allot them less land in land claims cases (Harris, 1998).

Loss of Land

During the early colonial period, European powers fought each other over land and resources. In North America, the British and French fought each other for the land mass we now know as Canada, with wars and skirmishes lasting from the early 1600s until the mid-1700s. The British defeat of the French in 1759 led to the signing of the Proclamation of 1763, which partially acknowledges Aboriginal territorial rights.[6]

In spite of the Proclamation and the promises it made, the British set about to remove First Nations people from their land and in the process colonize the entire country. On the one hand, the Proclamation recognized First Nations' rightful interest and use of the land, and stipulated that no First Nation land could be bought or taken without the negotiation of a treaty or the consent of the people in the area in question (Frideres, 2011, p. 9). On the other hand, what became increasingly important for the government was the issue of land acquisition—and

---

5. These trading posts also acted as conduits for European trade goods that First Nations were becoming increasingly dependent on through their direct involvement in the fur trade. Trading posts sold or traded muskets, which led to increasingly deadly conflicts between once-peaceful First Nations communities (Dickason & Newbigging, 2010; Harris, 1998).

6. The Royal Proclamation of 1763 has been referred to as an "Indian Bill of Rights" for its recognition of Aboriginal title, use of land, and the role that the Crown and federal governments had and continue to have in protecting Aboriginal interests. The Proclamation was passed in large measure to protect First Nations rights from the increasingly hostile treatment from "settlers" coming to Canada and encroaching on First Nations land and getting into physical altercations with them. Particularly important is how it has been interpreted to settle more recent land claims. It stipulates that Aboriginal land cannot simply be taken from them without the government first negotiating a treaty with the respective Aboriginal community. Settling current land claim disputes is often protracted and divisive as the Proclamation has set the precedent for all land claims.

for the government to secure it from First Nations communities.

Relevant Ideologies: Capitalism, Eurocentrism, and Patriarchy

The Commercialization of Relationships: The Fur Trade

The strategy of colonization was to increase wealth for the colonizer; hence, relations with First Nations were commercialized, which resulted in their increased dependency on European trade goods. The impetus for this activity was economic imperialism, and it would happen through the exploitation of Canada's many natural resources, many of which were very familiar to Aboriginal peoples. This of course led to their employ with various private and Crown corporations, such as the Hudson's Bay Company. Numerous laws were passed that sought to permanently alter Aboriginal ways of life, restricting where they could live, the customs they could practise, their freedom of movement and assembly, and introducing a patriarchal system of governance and decision making.

The fur trade altered the land in ways that would also permanently alter Aboriginal ways of living. Over-harvesting for the commercial trade led to a gradual decrease in wildlife. This was a condition that Aboriginals had never encountered. Their inability to deal with this new reality also created new problems, among which was the replacement of traditional materials and resources with imported European goods.

Also, traditional practices around hunting for the tribe became affected. Hunters now left the tribe for long periods to engage in commercial hunting, thereby effectively engaging in a form of wage-based labour, a concept completely foreign to them before Europeans arrived. With the decrease in the animal pelts that were coveted by the fur trade, it also meant increased conflict between different Nations as they competed for dwindling resources on each other's lands. Eventually, animal populations were so reduced that many Aboriginal communities were left destitute. Since they could not trap, they could not eat either (Ray, 1998, p. 97).

Upon European arrival, colonizing techniques such as pitting various First Nations against each other, otherwise known as "divide and conquer," were used. For example, trade wars began between various Aboriginal communities because they wanted access to the unusual trade goods (metal objects such as axes, awls, knives, glass beads, and cloth). Those conflicts first occurred in what is now Newfoundland among the once peaceful Beothuk, Mi'kmaq, and Abenaki. The consequence of those trade wars resulted in the extermination of the Beothuk, with the last one dying of tuberculosis in 1829 (Dickason & Newbigging, 2010; Upton, 1977).

Instruments of Control: Patriarchy

Colonialism and a patriarchal ideology resulted in the breakdown of egalitarian systems of governance, systems that had defined First Nations philosophy for millennia. Legislative changes introduced by the colonial government had a notable impact on Aboriginal women's rights. A patriarchal system of governance and family structure was introduced into First Nations communities.

As part of the emerging Indian Act, the 1857 Act to Encourage the Gradual Civilization of the Indian Tribes was passed by the colonial government. It marks a legislative start in the discrimination against Aboriginal women by allowing only males the right to vote (admittedly, no women of any race could vote in Canada until 1916), and successful male applicants would also be given 20 hectares of taxable reserve land (seen as another assimilationist tactic).

Perhaps even more of an affront to Aboriginal women's dignity and rights was the passing of the 1869 Act for the Gradual Enfranchisement of Indians. The act stipulated that Aboriginal women would lose their "Indian" status if they married a non-Aboriginal man, but if an Aboriginal man married a non-Aboriginal woman he would not lose his status and his non-Aboriginal wife would gain "Indian" status. The act has been seen as an effort to destroy Aboriginal ways of life.

It was not until decades later that this act was struck down as unconstitutional,[7] after the battle for Aboriginal women's rights gained speed in the 1950s. Many Aboriginal people challenged the law that stipulated that a First Nations woman and her children would be stripped of their status and forced to leave the reserve if she married a non-First Nations man (Dickason & Newbigging, 2010).

Structures of Domination: Reserves, Treaties, the Indian Act, and Residential Schools

Permanent changes to First Nations ways of life were brought about through assimilationist policies and laws introduced by the colonial and Canadian federal governments. Every advancement, modification, and adjustment to law or policy directed at First Nations was meant to reduce their standing as sovereign beings and to advance their forced removal from their land and the subsequent appropriation of those lands (Marule, 1981, p. 14).

Disenfranchisement from the Land: Reserves and Treaties as Instruments of Control

First Nations endured many challenges as a result of colonization. The imposition of immigration coupled with the relentless pace at which their land was seized by the colonial government was significant. [8] During the same period (1820–1850), the government was restricting Aboriginal movement in an attempt to isolate them from the growing European populations. The formation of the reserve system, the imposition of foreign political practices such as the forced adoption of an electoral system for electing chiefs, the banning of many traditional cultural practices, and the Indian Act, were encroaching on the Aboriginal life as they knew it.

For example, cultural practices such as the potlatch, a custom unique to the west coast First Nations of British Columbia where individuals would give away all of their possessions, was banned. Also, dances associated with traditional religious and supernatural beliefs and practices, such as the Sun Dance and the Ghost Dance of the plains First Nations, were outlawed by 1895 as they were seen as "paganistic" and an affront to Christianity (Dickason & Newbigging, 2010, p. 199; Miller, 2009, pp. 232, 247).

For assimilation and disenfranchisement to work, First Nations had to suffer loss of sovereignty through loss of land and gradual loss of self-government. This happened through their forced migration onto reserves, which thereby created a dependence on the state for food rations and other necessities. One of the greatest inequities faced by First Nations communities was the destruction of their own laws and customs, which were replaced by the laws of the colonialists. And while this happened over a time of more than 130 years, some of the bigger changes came with the enactment of the Canadian Constitution in 1867 (previously named the British North America Act, 1867), the signing of Treaties 1 through 6, the denial of treaties in some parts of the country, and the establishment of reserves (Miller, 2009, pp. 142–44).

Treaties: Legislated Theft

The treaties that were negotiated between the federal government and various First Nations of the plains (early 1870s) were drafted in response to the threat of First

---

7. In 1981 Sandra Lovelace, a Maliseet woman from the Tobique reserve in New Brunswick, wrote to the United Nations Human Rights Committee complaining of the discriminatory treatment faced by Aboriginal women. The Canadian government was forced to amend the legislation, which would thereby allow Aboriginal women the right to marry who they wanted, and keep their status and other provisions (Long & Dickason, 2011, p. 259).

8. The colonial government under Great Britain, and later the newly minted Canadian government (in 1867), were constantly under the threat of annexation by the United States. Many provinces were not sure that they wanted to join confederation as they had set up lucrative trade relations with their southern neighbours. Many in the U.S.A., including the government, believed in a "manifest destiny" in which they would have control of the entire continent. Hence, the goals of the Canadian government were to populate the country as quickly as possible and to have total control over First Nations so "settlers" would be able to take up their place in the new dominion.

Nations uprisings.[9] The federal government saw treaties as a way to manage native peoples and pave the way for widespread European settlement. The Plains Cree in particular fought for assurances that their autonomy, including where they could live, would be maintained if treaties were signed. They also petitioned the federal government to limit the buffalo hunt to First Nations only. This proved to be very difficult as not only were the buffalo herds dwindling rapidly, but the Crow, Peigan, and Blackfoot First Nations on both the American and Canadian sides of the border were also asking for restrictions—and the Cree were not originally from that area but had migrated in search of a reliable food source (Tobias, 1998).

The presence of First Nations people on a land that was destined for colonization was labelled as the "Indian problem," a term used as far back as the 1820s. For the British colonial governments and immigrants alike, the "Indian problem" meant that First Nations stood in the way of colonization. It was even believed that Aboriginal people were disappearing (dying off), and in a sense they were as their numbers had dropped by roughly 6000 in Upper and Lower Canada (modern day Ontario and Quebec) by 1830 (Dickason & Newbigging, 2010).

The establishment of reserves contributed to the oppression of First Nations people. Once on the reserves, regardless of whether they wanted to engage in agriculture (which is what the federal government said they had to do) they were all but abandoned, with most treaty promises being broken (Frideres, 2011). "The reserves were the cradle of the Indian civilizing effort—and the means of securing the White man's freedom to exploit the vast riches of a young dominion" (Dickason & Newbigging, 2010, p. 171).

The federal government looked at the treaties and reserves as bestowing a privilege, when in fact they had totally altered First Nations' ways of life and subjugated them. First Nations who were dependent on the buffalo hunt for food and subsistence went from prosperity to begging for food from the white settlers who took their land (Dickason & Newbigging, 2010).

Residential Schools: Removing the Indian from the Child

As much as the treaties and the reserve system had a devastating effect on First Nations, the residential school system dealt a particularly harsh blow. The trauma of residential schools continues to affect the Aboriginal community today. Its purpose was to extinguish any traces of Aboriginal culture and identity.

The residential school system began at various times across the country: in Quebec in the early 1700s (Barman, Hebert, & McCaskin, 1986, p. 4), and a hundred years later in the Prairies. It was not always controlled by the colonial or federal government and did not become a central pillar of the Indian Act until the mid-1800s. The colonial, and later the federal, government decided that rather than start its own system, it would work with the churches that had started schools in an effort to Christianize First Nations.

After the completion of the continental railroad in 1885, demands grew for the complete assimilation of First Nations people, and the residential school system took up a central role in that goal. The Canadian government decided that the best way to educate (assimilate) First Nations children was to remove them from their homes and place them in residential schools. In most cases, children had no contact with their families. All aspects of their lives, including language, clothing, hairstyles, and conduct, would be strictly controlled through corporal punishment and modelled after European expectations.

---

9. By the 1880s, the Cree were the most militant First Nations group in Canada, and the federal government expressed concerns over their number and their increasing demands for land and mobility rights. Their distrust of the government coupled with outright anger over broken promises led to one of the largest Aboriginal uprisings in Canadian history in 1885. Métis leader Louis Riel joined forces with Cree leader Big Bear and stormed the garrison at Batoche, Saskatchewan, in an attempt to create a Métis and First Nations homeland. The North West Rebellion, as it is known, was put down, Riel was hung for treason, Big Bear was imprisoned and later released, and several other Métis and First Nations people were also hung or imprisoned (Dickason & Newbigging, 2010, p. 218). The message was clear: there would be no territory for Aboriginal people, and the government would use whatever force necessary to meet its goals.

It has been suggested that the conditions at many residential schools where children were boarded was so poor in the late 1800s to early 1900s that it would be generous to say that 50 percent of them lived to see the benefits of the education that they were forced to endure (Barman et al., p. 8). Another startling effect of this system was that after years away from their families, when the children returned to the reserves at age 16, not only could they not fit back into their own communities, but they weren't welcomed in the white world either (Barman et al., p. 9). In effect, they were left stateless in their own land. To make matters worse, the quality of their education was below that of the schools white children went to and when they finished (in the early 1900s) they weren't allowed to go to high school (Barman et al., 1986, p. 11), thereby ensuring their continued second-class-citizen status.

However, First Nations did not respond passively. Many resisted and refused outright to send their children away. This wasn't just because of the poor quality of the education itself, but because they were aware of the inherent contradictions in the educational policy itself that continued to discriminate against them even if they went to school. By the mid-1900s, in what is now referred to as the "baby boom" generation, only a third of First Nations children went beyond Grade 3 compared to over 60 percent in the white population (Barman et al., 1986, p. 13). It wasn't until after World War II that the federal government was forced to recognize the inadequacies of their education policies towards First Nations (Perrson, 1986).

Truth and Reconciliation

In 2008, more than 40 years after the closing of the last residential school, First Nations finally received an apology from the federal government of Canada for the abuse they suffered in the residential school system. The apology was seen as a milestone in federal–First Nations relations. It represented the first time that the federal government actually used the words "we're sorry," which was taken as an admission of responsibility for the damage caused to so many First Nations families. The Residential School Settlement Agreement of 2007 compensated roughly 80,000 survivors of the residential school system, and also led to the creation of the Indian Residential School Truth and Reconciliation Commission (Frideres, 2011; Regan, 2010).

The reconciliation process had been occurring since the early 1990s when the various churches involved in administering residential schools started apologizing to First Nations communities for the physical, sexual, and mental abuse inflicted upon their children by clergy. However, it was only during the federally appointed Royal Commission on Aboriginal Peoples[10] that survivors of the residential school system really started telling their stories of abuse, of the deep emotional scars that remained, and of how the pain caused by the abuse has created intergenerational family problems. Aside from the apologies and payouts made by the various churches, in 1998 the federal government under Paul Martin offered $350 million to aid in the recovery of victims of the residential school system.[11] The money that was offered was seen as a first step and not nearly enough to compensate the thousands of survivors. Under increasing pressure from First Nations survivors and

---

10. In 1996, the federal government commissioned the 4000-page Royal Commission on Aboriginal Peoples report. The commission made over 400 recommendations regarding issues that characterize life on many reserves across Canada (such as the lack of proper sanitation and sewage systems resulting in boil water advisories; tuberculosis; high rates of suicide, unemployment, and incarceration compared to the general population; decrepit infrastructure; and very low educational retention rates). Almost 20 years later, the recommendations mostly remain unfulfilled. In 2006, the CBC, and in 2010 the Toronto Star, conducted a series of investigations into the living conditions, financial problems, and other issues plaguing reserves across Canada. They found that 117 reserves, with a population of over half a million people, had been under boil water advisories for years.

11. In order to help the churches avoid bankruptcy, the federal government stepped in with this initial compensation. The churches were losing a growing number of lawsuits and could not afford the amount of compensation awarded to victims.

legal representatives, the federal government was eventually persuaded to act. In May 2006, a deal was reached, and the largest class action law suit in Canadian history awarded First Nations survivors $1.9 billion through what came to be known as The Indian Residential School Settlement Agreement (Tailfeathers, 2012, p. 5).

The Failure of the Education System and the Case for Self-Government

The inadequacies of the education system imposed upon First Nations, combined with their awareness of this discriminatory practice and of the emergence of various human rights movements in the 1960s, helped to pave the way for the idea of self-government. First Nations maintained that the colonial education system was not working and that education for First Nations should be handled by First Nations.

In 1969, the federal government under Prime Minister Pierre Trudeau released a White Paper that spelled out how dismantling the Indian Act and the reserve system and severing all ties with First Nations would give them the same chances as their non-Aboriginal counterparts. First Nations groups across the country, such as the National Indian Brotherhood (now known as the Assembly of First Nations), reacted swiftly (Miller, 2009, p. 248). Trudeau was forced to back down, signalling an end to the cultural assimilationist approach that had defined colonial/federal government–First Nations relationships for the previous two centuries.

Self-Government

One way to understand self-government is to see it as a shift in power from being controlled by a foreign, external entity (the federal government and the Indian Act) to First Nations having control over their education, justice (certain aspects of it), lands, resources on those lands, health and social services (Barman et al., 1986). In short, self-government is the ability of a group of people to govern themselves and their lands without having to ask permission to do so (Asch, 2002, p. 66). First Nations arguments have been consistent; they were self-governing people before European colonizers arrived. Rights to the land and its resources, their culture, languages, customs, and sovereignty cannot be conferred upon them by a late-coming benevolent dictator.

First Nations' quest for self-government is not just a fight against the federal and provincial governments. It is also a fight against popular perceptions of Aboriginals held by the majority of non-Aboriginals, and against corporations, particularly those involved in mineral and resource exploration and exploitation. However, self-government does not mean extinguishing business or constitutional relations between the federal government and First Nations; it is precisely the federal government that needs to enshrine those responsibilities in law (Marule, 1981, p. 2).

First Nations believe that the Constitution, the Charter of Rights and Freedoms, and all of the previous treaties and conventions between First Nations and the federal government contain the foundation for the realization of First Nations' goals of self-government.[12] The current Constitution (1982)[13] does recognize the collective interests of First Nations and also recognizes that they have exclusive rights—indeed, a proprietary interest—to all lands reserved for First Nations (Union of B.C. Indian Chiefs, 1980, p. 2). Also, it is expressly stated in the Royal Proclamation, 1763, through the Doctrine

---

12. The Union of B.C Indian Chiefs identify the following examples: the Treaty of Utrecht, 1763; the Treaties of Peace and Friendship with the Indian Nations of Nova Scotia and New Brunswick, 1713 and 1763; the Articles of Capitulation of Quebec, 1757 and of Montreal 1760 and 1763; the Treaty of Paris 1763; the Royal Proclamation, 1763; and the Constitution Act, 1791.

13 Working against the attempt to realize self-government are the repeated attempts by the federal government to circumvent or outright extinguish Aboriginal title (Marule, 1981). The repatriation of the Constitution from Great Britain in 1982 by Prime Minister Pierre Trudeau was no different as the federal government once again tried to extinguish First Nation title through amendments to the Constitution. Although provincial premiers were invited to the negotiations on the amendments, First Nation political leaders were not. Their exclusion was one reason they objected to the repatriation and successfully forced the deletion of amendments that would have seen the federal government absolve itself of Constitutional responsibilities (Union of B.C. Indian Chiefs, 1980, p. 14).

of Consent, that the (British) Crown "would recognize the lawful transfer of Indian lands to the Crown if, and only if, the Indian Nations gave their consent" (Marule, 1981, p. 3). Despite these legislative assurances, the struggle for self-government has been paved with difficulties.

One of the main goals in the quest for self-government has been for First Nations to hold the Crown and the federal government accountable for the treaties and various political declarations that have been signed (Marule, 1981, p. 2). Many are versed in the legalese of the government and know that their rights cannot be so easily extinguished without a complete repudiation by the Crown and federal government of their own conventions and laws. In other words, they state that First Nations "will never be legislated out of existence" (Union of B.C. Indian Chiefs, 1980, p. 21).

At the international level, the United Nations Covenant on Economic, Social and Cultural Rights, Article 1, Part 1, states that "All peoples have the right of self-determination. By virtue of that right they freely determine their political status and freely pursue their economic, social and cultural development" (Marule, 1981, p. 16). Despite being a party to this international treaty since 1976, the Canadian government has repeatedly denied First Nations the right to self-determination (that is, self-government).

First Nations have identified specific recommendations to bring self-government to fruition. Due to their decentralized nature across the country, self-governing First Nations would not look like provinces. Rather, through a constitutional process, a First Nations legislative assembly would be the governing body for First Nations people, and it would have the powers and resources of a provincial government, as well the rights as spelled out in the treaties, such as fishing and hunting. It would also have powers over who can have membership in a Band, the administration of cultural, social, education, health and other services, and the administration of equalization payments that provinces are entitled to (Union of B.C. Indian Chiefs, 1980).

First Nations argue that the Indian Act and the other statutes should be answerable to the Charter of Rights and Freedoms, which in theory supersedes all other legislation. However, as Marule (1981), the B.C. Union of Indian Chiefs (1980), Flanagan, Alcantra, and Le Dressay (2011), and Dickason and Newbigging (2010) point out, the Indian Act continues to define Aboriginal rights, keeps Aboriginals separate from the rest of the population, and that assimilation has emphasized control over their improvement.

Today's Indigenous Resistance to Colonialism

Perhaps the biggest Aboriginal movement today is Idle No More. While it was started in Canada in 2012 by four First Nations women from Saskatchewan, it has grown internationally as a vehicle to unite Aboriginal concerns for self-government and environmental sustainability. They are on the front lines of pushing for decolonization.

Idle No More asserts that non-Aboriginal Canadians also benefit from the disenfranchisement of Aboriginals, but do so at everyone's peril as the skyrocketing costs of incarceration, poor health, and exceedingly low post-secondary graduation rates for First Nations end up costing all Canadian taxpayers. They further argue that the destruction of Canada's natural environment as a result of resource exploitation concerns all Canadians too.

Through protests, speaking engagements, and publicity campaigns, they are aggressively pushing for Aboriginal sovereignty in Canada and raising awareness of environmental issues, particularly the Alberta oil sands and the environmental destruction underway there (Donkin, 2013).

Idle No More has also received considerable media coverage. One of their goals is to inform the public about the misdeeds and misinformation of the federal government. They point out that the federal government continues to take an assimilationist approach to First Nations. They suggest that the reason federal funding for many Aboriginal communities is so low is that the government is trying to encourage Aboriginal people to move off the reserve and into the cities, effectively being assimilated.

CONCLUSION

This chapter has attempted to offer the reader a glimpse into some of the historical and contemporary issues around the relationship of Canada and its First Nations. The events of the past few hundred years are far more complicated than what has been discussed, but the discussion nonetheless offers insight into some of the main themes and circumstances facing First Nations.

First Nations continue to live under the Indian Act. While there have been certain gains in self-government, such as the Nishga'a land claims settlement and the creation of Nunavut, much more work needs to be done. Aboriginals have the highest incarceration rates in Canada. While representing only 4 percent of the population, they make up almost 24 percent of the prison population (Government of Canada, 2014). Education and housing issues continue to plague First Nations communities with 36 percent of First Nations youth graduating from high school compared to 72 percent for the rest of the population. Suicide among Aboriginal youth is 5 to 7 times the national average (Assembly of First Nations, 2012). On many First Nations reserves, boil water advisories are in effect due to water pollution (as along the MacKenzie River) and poor sanitation infrastructure. The failure of successive federal governments to adequately address these issues is one of the primary reasons behind the First Nations quest for self-government.

CRITICAL THINKING QUESTIONS

1. How does the concept of equity play a role in hiring practices?

2. What is the relationship between poverty and other social inequalities? Why does socio-economics play such an important role in terms of its contribution to inequality?

3. Discuss Canada's image as an egalitarian multicultural society internationally as it relates to global citizenship.

4. Discuss the complexities of power and privilege as they relate to different aspects of our identity. How does this translate to our position in society and our experiences of inequality?

REFERENCES

Adams, M., Bell, L. A., & Griffin, P. (1997) *Teaching for diversity and social justice.* New York: Routledge.

Adelman, H. S. (1992). The classification problem. In W. Stainback & S. Stainback, *Controversial issues confronting special education: Divergent perspectives.* London: Alyyn and Bacon.

Anzovino, T., & Boutilier, D. (2015). *Walk a mile: Experiencing and understanding diversity in Canada.* Toronto: Nelson Education.

Asch, M. (2002). Self-government in the new millenium. In J. Bird, L. Land, & M. McAdam (Eds.), *Nation to nation. Aboriginal sovereignty and the future of Canada.* Toronto: Irwin Publishing.

Assembly of First Nations. (2012, October). A portrait of First Nations education. Retrieved from http://www.afn.ca/uploads/files/events/fact_sheet-ccoe-3.pdf

Axtel, J. (1998). Through another glass darkly: Early Indian views of Europeans. In K. S. Coates & R. Fisher (Eds.), *Out of the background. Readings on Canadian Native history* (2nd ed.). Scarborough, ON: Irwin Publishing, Thomson Nelson.

Barman, J., Hebert, Y., & McCaskill, D. (1986). The legacy of the past: An overview. In *Indian education in Canada: The Legacy* (Vol 1). Vancouver: UBC Press.

Bell, L. A. (1997). Theoretical foundations for social justice. In M. Adams, L. A. Bell, & P. Griffin, P. (1997) *Teaching for diversity and social justice.* New York: Routledge.

Bottero, W. (2004). *Stratification: Social division and inequality.* Florence, KY: Routledge Taylor and Francis Group.

CBC. (2013, September 10). Charter of Quebec values would ban religious symbols for public workers. Retrieved from http://www.cbc.ca/news/canada/ montreal/charter-of-quebec-values-would-ban-religious-symbols-for-public-workers-1.1699315

Curtis, J., Grabb, E., & Guppy, N. (2004). *Social inequality in Canada: Patterns, problems, and policies* (4th ed.). Toronto: Pearson Publishers.

Dickason, O. P., & Newbigging, W. (2010). *A concise history of Canada's First Nations* (2nd ed.). Don Mills, ON: Oxford University Press.

Donkin, K. (2013, January 18). Idle No More: The rise of Pam Palmater. *The Toronto Star.* Retrieved from http://www.thestar.com/news/weekend/2013/01/19/idle_no_more_the_rise_of_pam_palmater.html

Flanagan, T., Alcantara, C., & Le Dressay, A. (2011). *Beyond the Indian Act. Restoring Aboriginal property rights* (2nd ed.). Montreal and Kingston: McGill-Queen's University Press.

Fleras, A. (2012). *Unequal relations: An introduction to race, ethnic, and Aboriginal dynamics in Canada* (7th ed.). Toronto: Pearson Publishers.

Fraser, J. (2012). Canada's income inequality: What is it, and how bad? Huffington Post. Retrieved from http://www.huffingtonpost.ca/2012/05/27/canada-income-inequality-by-numbers_n_1545900.html

Frideres, J. S. (2011). *First Nations in the twenty-first century.* Don Mills, ON: Oxford University Press.

Gibson, A. M. (1980). *The American Indian. Prehistory and present.* USA: D.C. Heath and Company.

Gilbert, D., & Kahl, J. A. (1998). *The American class structure: A new synthesis* (5th ed.). Belmont, CA: Wadsworth.

Government of Canada. (2014). Office of the Correction Investigator. Backgrounder. Aboriginal offenders—A critical situation. Retrieved from http://www.oci-bec.gc.ca/cnt/rpt/oth-aut/oth-aut20121022info-eng.aspx

Grabb, E. (2004). Conceptual issues in the study of social inequality. In J. Curtis, E. Grabb, & N. Guppy, *Social inequality in Canada: Patterns, problems, and policies* (4th ed.). Toronto: Pearson Publishers.

Harper, D. J. (2011). Social inequality and the diagnosis of paranoia. *Health Sociology Review, 20*(4), 423–36.

Harris, C. (1998). Voice of disaster: Smallpox around the Strait of Georgia in 1782. In K. S. Coates & R. Fisher (Eds.), *Out of the background. Readings in Canadian Native history* (2nd ed.). Scarborough, ON: Irwin Publishing.

Hulchanski, D. (2010). The three cities within Toronto: Income polarization among Toronto's neighbourhoods. Retrieved from http://www.urbancentre.utoronto.ca/pdfs/curp/tnrn/ Three-Cities-Within-Toronto-2010-Final.pdf

Jianghe, N., & Rosenthal, S. A. (2009). Trust discrimination toward socially dominant and subordinate social groups. *North American Journal of Psychology, 11*(3), 501–09.

Jordan, A. (2007) *Introduction to inclusive education.* Mississauga, ON: John Wiley & Sons Ltd.

Kendall, D. (2010) *Sociology in our times: The essentials.* Belmont, CA: Wadsworth.

Kendall, D., Lothian-Murray, J., & Linden, R. (2007). *Sociology in our times.* Toronto: Nelson Education.

Lilly, S. M. (1992) Labeling: A tired, overworked, yet unresolved issue in special education. In W. Stainback & S. Stainback, S., *Controversial issues confronting special education: Divergent perspectives.* London: Alyyn and Bacon.

Long, D., & Dickason, O. P. (2011). *Visions of the heart. Canadian Aboriginal issues* (3rd ed.). Toronto: Oxford University Press.

Lopes, T., & Thomas, B. (2006). *Dancing on live embers: Challenging racism in institutions.* Toronto: Between the Lines Publishing.

Macdonald, D. (2014). Outrageous fortune: Documenting Canada's wealth gap. Canadian Centre for Policy Alternatives. Retrieved from https://www.policyalternatives.ca/sites/ default/files/uploads/publications/National%20Office/2014/04/Outrageous_Fortune.pdf

MacKinnon, C. (1989). *Equality rights: An overview of equality theories.* Ottawa: National Meeting of Equality Seeking Groups. In T. Anzovino & D. Boutilier (2015), *Walk a mile: Experiencing and understanding diversity in Canada.* Toronto: Nelson Education.

Marule, M. S. (Ed.). (1981). First Nations, states of Canada & United Kingdom: Patriation of the Canadian Constitution. Unpublished manuscript, prepared for Constitutional Committee of the Chiefs of Alberta.

McGregor, H. E. (2010). *Inuit education and schools in the eastern Arctic.* Vancouver: UBC Press.

McPherson, K. (2013). Moving to equity and advancing equality. In *GNED500 Global Citizenship: From Social Analysis to Social Action.* Toronto: Pearson Publishers Custom.

Miller, J. R. (2009). *Compact, contract, covenant. Aboriginal treating-making in Canada.* University of Toronto Press. Toronto.

Minister of Justice. (2003). *Canadian Multiculturalism Act.* Government of Canada. Retrieved from http://laws.justice.gc.ca/PDF/C-18.7.pdf

Ministry of Economic Development, Trade, and Employment. (2013). Understanding barriers to accessibility. Retrieved from http://www.mcss.gov.on.ca/en/mcss/programs/accessibility/ understanding_accessibility/understanding_barriers.aspx

Persson, D. (1986). The changing experience of Indian residential schooling: Blue quills, 1931–1970. In *Indian education in Canada: The Legacy* (Vol 1). Vancouver: UBC Press.

Ray, A. (1998). Periodic shortages, Native welfare, and the Hudson's Bay Company 1670–1930. In *Out of the background. Readings on Canadian native history* (2nd ed.). Scarborough, ON: Irwin Publishing, Thomson Nelson.

Regan, P. (2010). *Unsettling the settler within. Indian residential schools, truth telling, and reconciliation in Canada.* Vancouver: UBC Press.

Robbins, W. J. (2012). Discriminatory rationalization: The equity/excellence debate in Canada. Forum on Public Policy. Retrieved from http://forumonpublicpolicy.com/vol2012.no2/archive/robbins.pdf

Savage, M., Devine, F., Cunningham, N., Taylor, M., Li, Y., Hjellbrekke, J., Le Roux, B., Friedman, S., & Miles, A. (2013). A new model of social class: Findings from the BBC's Great British class survey experiment. *Sociology*, 1–32. doi: 10.1177/0038038513481128

Sensoy, O., & DiAngelo, R. (2012). *Is everyone really equal? An introduction to key concepts in social justice education.* New York: Teachers College Press.

Stainback, W., & Stainback, S. (1992) *Controversial issues confronting special education: Divergent perspectives.* Massachusetts: Allyn and Bacon.

Statistics Canada. (2011). *Immigration and ethnocultural diversity in Canada.* Government of Canada. National Household Survey. Retrieved from http://www12.statcan.gc.ca/nhs-enm/2011/as-sa/99-010-x/99-010-x2011001-eng.pdf

Statistics Canada. (2013). *Low income cut-offs.* Government of Canada. Retrieved from http://www.statcan.gc.ca/pub/75f0002m/2012002/lico-sfr-eng.htm

Statistics Canada. (2013). *Economic well-being. Women in Canada: A gender-based statistical report.* Retrieved from http://www.statcan.gc.ca/pub/89-503-x/2010001/article/11388-eng.htm

Tailfeathers, R. (2012, January). Niitsitapi—A resilient people. In Blood tribe chief and council review. *Tsinikssini, 4*(1). Retrieved from http://newsletters.bloodtribe.org/2012/JAN2012.pdf

Thobani, S. (2007) *Exalted subjects: Studies in the making of race and nation in Canada.* Toronto: University of Toronto Press.

Tobias, J. L. (1998). Canada's subjugation of the Plains Cree, 1879–1885. In *Out of the background. Readings on Canadian native history* (2nd ed.). Scarborough, ON: Irwin Publishing, Thomson Nelson.

Union of B.C. Indian Chiefs. (1980, October). Indians and the Canadian Constitution. A position paper. Vancouver: Union of B.C. Indian Chiefs.

United Way Toronto. (2004). *Poverty by postal code: The geography of neighbourhood poverty City of Toronto, 1981–2001.* Retrieved from http://www.unitedwaytoronto.com/downloads/whatWeDo/reports/PovertybyPostalCodeFinal.pdf

Upton, L. (1977). The extermination of the Beothucks of Newfoundland. *Canadian Historical Review, 58*, 144–53.

Zuriff, G. E. (1996, October). The myth of learning disabilities: The social construction of a disorder. *Public Affairs Quarterly, 10*(4), 276.

Chapter 8

Making A Difference Through Social Action

Paula Anderton with Rosina Agyepong

LEARNING OUTCOMES

LO-1 Define different levels of social action

LO-2 Distinguish between social action done by social justice movements versus extremist movements

LO-3 Describe four stages of social movements and explain the relationship between social movements and social change

LO-4 Analyze the connection between social action and social justice

LO-5 Evaluate the role of business and technology in social action

LO-6 Evaluate the role of the individual in social action

LO-7 Reflect on the importance of social action

> We don't have to engage in grand, heroic actions to participate in the process of change. Small acts, when multiplied by millions of people, can transform the world.
>
> —Howard Zinn

INTRODUCTION

Throughout this book, various political and cultural critiques have been raised about what it means to be a citizen in this interconnected world. This chapter will explore what global citizenship means in terms of social action, the point at which awareness is put into practice. When "this isn't fair" becomes "let's *make* it fair," the shift has been made from awareness to action. **Social action**, then, is any action of an individual or group who seeks to promote social change on a small or large scale. There is more to social action, however, than the willingness to get involved. Good intentions can sometimes do more harm than good. Some of the most vocal and committed individuals are working for extremist groups willing to kill to promote their values. This chapter focuses on social action in the service of social justice.

social action

Individual or group behaviour that involves interaction with other individuals or groups, especially organized action towards social reform. Social action is an action by a group of people directed toward a better societal end.

It distinguishes between individuals and groups seeking to relieve suffering and promote equity, and those promoting extremism. This chapter outlines the main approaches to social action, profiles some notable figures in social activism, and introduces some of the new thinking about communications technology as a tool for global change. Furthermore, a critical analysis is used in order to understand how factors such as ideology and privilege can complicate and even subvert our attempts to help one another. Ultimately, the reader will be asked to reflect on what constitutes productive social change and what that means to us as individuals.

WHY TAKE SOCIAL ACTION?

Nine-year old Martha Payne didn't like her lunch—the ones served by her school. In 2012, she started a blog that documented the unhealthy and unappealing food given to students in her school in Argyll, Scotland. Her posts quickly went viral, receiving 100,000 visitors in the first week, even attracting the support of chef-**activist** Jamie Oliver.[1] Martha's criticism of her school lunches sparked debate about school nutrition and freedom of expression in the United Kingdom and the issue quickly became a headline on the Internet.

activist
An individual who devotes time to work either paid or unpaid to bring about social change.

Why take social action? Because a social condition is unacceptable and requires change. Social action can start with the simplest idea and the willingness to implement it. Martha Payne could not have predicted that her critique on a local issue would spark debate around the world about the food industry, education, and even child hunger. Doubtless, Martha wasn't the only student who complained about her lunch, but she did something about it. Martha demonstrated that thoughtful critique, when given structure and a goal, can be effective social action. Her blog resonated with people all over the world and surfaced an underlying social problem: profit versus nutrition. Martha Payne demonstrated how one concerned individual with a desire for change can inspire collective action with significant results. The motives for taking action, however, are not always as straightforward as Martha's desire for improved school lunches.

---

1. British chef Jamie Oliver, well known for his cooking shows, cookbooks, and kitchen products, has become widely recognized for his social actions. His not-for-profit restaurant, Fifteen, gives marginalized youths commercial kitchen training and life skills. His activism on the rising rates of obesity in England, initiated his high-profile campaign to reform eating habits throughout the industrialized world, starting with children. His televised campaign to eradicate processed food from British and U.S. school cafeterias has led to several successful initiatives to help people eat well and sustainably. It is through this work that he became aware of Martha Payne and publicly supported her cause.

SOCIAL ACTION AND BACKLASH

Social action is widely understood as action that promotes a goal of positive change. However, not all social action pursues change in the direction of equity and social justice. In fact, social action in pursuit of progressive values such as equity often generates a backlash from groups who are threatened by such change, either through loss of privilege or a perceived assault on their identity. Think of the actions and rhetoric of militant fundamentalist groups reacting to progressive values such as equality for women, birth control, or freedom of expression. Malala Yousafzai[2] was the victim of such a backlash in Pakistan. The social action taken by these groups is very different in goal and approach from the variety that pursues social justice.

Strictly speaking, social action can be initiated by anyone, with any agenda. It could be political, social, or community groups serving a variety of interests that may be contrary to the principles of social justice. Hate groups, for instance, are some of the most active and effective users of both local organizing and global communications. For example, if you follow a link at the bottom of the website, martinlutherking.org, you will discover that the website is hosted by a self-described white nationalist group called Storm Front, and its purpose is to defame Dr. King and the work he did to advance civil rights for African-Americans. In Mangalore, India, another xenophobic group called Sri Ram Sene organizes assaults on Hindu women whom they have judged to be in moral violation of their fundamentalist beliefs. These groups are using the methods of social action to marginalize, control, and terrorize people who do not share their ideology. They are, in fact, misusing social action. Their goals are oppressive and require force to maintain them. This approach fundamentally negates society.

Maajid Nawaz, a vocal anti-extremism advocate, comments on this type of social action by saying, "Ironically, xenophobic nationalists are utilizing the benefits of globalization." He describes in detail how they are successfully spreading their agenda both on the ground and online, using the borderless tools of the Internet (Nawaz, 2011). In a globalized world, the publicity these hate groups generate attracts critics as well as new supporters. Internet exposure can often swell the ranks

---

2. At age 15, Malala Yousafzai, was shot and almost killed in 2012 by Taliban extremists in Pakistan who objected to her vocal defense of the right of girls to receive education. The shocking nature of the attack, coupled with the bravery of her continued resistance to the oppression of girls after her recovery, has made her an international figure. She has received several high profile humanitarian awards, including the UN Human Rights Prize, a distinction she shares with the late Nelson Mandela. She was also nominated for the 2013 Nobel Peace Prize and has a foundation named after her. The Malala Fund, as it is called, actively promotes the universal right to education and includes support from the UN Foundation, among other high-profile aid organizations. The Taliban backlash against her activism failed to silence her and has in fact promoted her cause.

of small, localized extremist groups by reaching out to like-minded individuals. Ultra-conservative groups that may have anti-immigration, **misogynist**, racist, or anti-equity agendas are highly motivated in their pursuit of social action and recognize the value of new technology in recruiting followers. Indeed, Nawaaz (2011) argues that this willingness of ideologically motivated hate groups and extremists to take action using every means available is precisely what makes them dangerous but also effective. It is for this reason that social action of this type—extremist, oppressive, violent—is given space in this chapter. It is not genuine social action, because it is fundamentally antisocial and promotes the interests of the few, not the many. However, it must be recognized in order to be resisted. Social action to promote social justice must counteract social action for oppression with equal or greater dedication, argues Nawaz. If not, social action will be used to make the world less equitable and hard won gains for social justice could be lost. This cause is more urgent than ever in the globalized, interconnected world where extremism can be promoted and organized through technology. The social justice backlash phenomenon must be acknowledged not to legitimize it, but to eliminate it.

FOUR MODELS OF SOCIAL ACTION

Martha Payne turned her individual effort into meaningful social action, which resulted in increased awareness, public debate, and improved lunches in her school—remarkable outcomes from a modest undertaking. Clearly the actions of individuals can be powerful, but this is just one example of many types of social action. Broadly speaking, most social action can be characterized as bottom-up or top-down approaches to solving social problems. Top-down solutions are institutional in scale such as **international aid** and large **charities,** which will be discussed in the following section. Bottom-up solutions, also known as **grassroots** actions, are typically small-scale, locally based efforts to fix a problem or change a policy. The examples of Barefoot College and the Occupy Movement will illustrate this approach and some of the issues surrounding it. The third-model examined will be **online activism**. It has many of the criteria of grassroots activism and it is driven by individuals often acting alone in front of their computers. Yet, it relies on large, globalized corporate communication networks, blurring the standard ideological lines that traditionally separate big business and grassroots activism. In other words, it is a new hybrid social action that uses top-down infrastructure for bottom-up activism. The final model in this section, social entrepreneurship, applies business principles to solving social problems and, like online activism, is generating fresh debate on what constitutes effective social action and who can conduct it.

misogynistic
A person or organization that hates, distrusts, or mistreats women.

international aid
(also known as foreign aid, or development) The transfer of resources—money, goods, expertise—from a country or large organization to a recipient country in order to help them emerge from poverty.

charity
Aid given to those in need. This could be done on an individual basis or by an institution or organization engaged in relief services for the poor and underprivileged in society.

grassroots organizations
Social activism at a local or community level. Grassroots organizations tend to work through existing political structures to promote social change.

online activism
Also known as cyberactivism or digital activism, uses Internet-based communication techniques to manage and promote activism of any type.

Model 1: Social Action From the Top Down—Big Aid

Social action that crosses borders is known as international development. When this development process involves countries and governments, it is known as foreign aid or simply, aid. Development is the level where globalization meets social action on a massive economic scale. It involves a variety of players from non-governmental organizations, who rely on donations or sponsorships, to international diplomatic structures and the global banking system. This level of aid has been controversial and raises questions about power, privilege, and the ideology of development itself.

Development in its present form is a concept created in 1945 at the end of World War II.[3] Prior to this, colonizing countries only pulled resources out of their acquired territories and any investment made in those countries supported this outflow of wealth.[4] No resources were ever brought in with the intention of benefiting indigenous populations until the end of the 1930s, an era sometimes called the "second colonial occupation." Colonial slave labour practices had created an impoverished, dependent, and rebellious population in colonized countries while undermining the social fabric and identity of those societies. This realization was finally being called into question from both moral and economic standpoints.[5] By the end of World War II, these colonized states were ready to reclaim their self-determination and identity, by force if necessary. The first development loans to these struggling, newly independent states were from their former colonizers who sought to maintain economic, if not territorial, domination in their former colonies. The post-war era saw the collapse of European colonialism coupled with the destruction of the war and the moral horror of the Holocaust. Against this backdrop, big aid was formed as a concept and the idea of nations helping other nations develop was characterized as a new paradigm in human relations. The reality has been somewhat less idealistic.

The period from 1945 to 1949 saw the birth of the United Nations, the International Monetary Fund (IMF), and the International Bank for Reconstruction and Development (World Bank). These institutions form the basis of the modern, globalized economic system we have today. Together with the U.S.-funded Mar-

---

3. Although the term "development" wasn't used until the post-WWII period, it is ideologically rooted in Enlightenment notions of human progress that have formed the rationale for free market capitalism (see Chapter 4). This western idea of human progress through economic growth has become the dominant discourse in development, ignoring other ways of viewing progress, or even the validity of the idea itself for the cultures on which it is imposed.

4. The Indian Famine of 1899 resulted in over 1 million Indians dying from starvation and accompanying disease. At the same time, Indian-produced grain was being exported to Britain.

5. See Walter Rodney's argument on how Europe's economic exploitation of Africa in combination with post-war power politics served to destabilize and handicap African development for decades, and continues to do so in an ongoing process of economic colonization (Rodney, 1973).

shall Plan, the goal of the first international development initiatives was to rebuild the economies of Western Europe and Japan after the war and to stop the territorial expansion of the Soviet Union.

The IMF and the World Bank were formed with the stated goals of ensuring sustainable economic growth while promoting international trade and employment, and reducing poverty worldwide—thus solving the social problems that fuelled World War II. This would be achieved through a system of development grants and loans—money redistributed from richer countries to poorer ones in order to help them build their economies and join the international community. Unfortunately, this did not happen. In the decades following the creation of this system, wars, political instability, and economic and social inequity have continued to plague the planet. Development money is distributed inequitably and loans for profit from huge private banks keep struggling nations in a perpetual cycle of debt payments instead of growth. Social action on a grand scale, with developed countries aiding developing countries, has not fulfilled its post-war promise of creating a better world for everyone. Critical analysis reveals fundamental flaws in the design of the top-down development model.

Critically Thinking About Big Aid

The history of international development has followed a pattern that is tied to global politics, not utopian ideals. The United States, the dominant economic power since World War II, has donated aid and development resources most readily to countries whose governments support their interests—anti-communist regimes in the past, and more recently, allies in the "War on Terror." Other developed countries, including Canada, have followed this model of aid that protects their political and economic interests. Whatever gains this has produced in the world of geopolitics, few would claim that the results on ground level have been *as advertised* in the post-war rhetoric of development. The great failure of modern, top-down social action has been that so much international aid money has been embezzled, mismanaged, or simply wasted on poorly planned projects:

> The West spent $2.3 trillion on foreign aid over the last five decades and still had not managed to get 12-cent medicines to children to prevent half of all malaria deaths. The West spent $2.3 trillion and still had not managed to get $4 bed nets to poor families. The West spent $2.3 trillion and still had not managed to get $3 to each new mother to prevent five million child deaths. … it is heart-breaking that global society has evolved a highly efficient way to get entertainment to rich adults and children, while it can't get 12-cent medicines to dying poor children. (Easterly, 2006)

Both the quantity and the quality of aid to poorer nations has been inadequate to create social action that is effective or lasting. That means there has not been enough aid money and that the money that *is* donated, is distributed in a way that decreases its practical value to the recipient. This is a widely misunderstood phenomenon. Average citizens of wealthy countries look on with dismay as poor countries seem to get poorer, despite all the international aid they receive. Thus, the failure of international development efforts is often explained by corruption within recipient countries—a "blame the victim" approach. The reality is far more complex. Certainly there is systemic corruption among elected and unelected governments alike in the developing world. But asking some basic social analysis questions when it comes to the failure of aid (such as, whose interests are served by this system?) reveals much more going on than corrupt dictators lining their pockets. Consider the following assertions from Pekka Hirvonen at Global Policy Forum:

- Although the Cold War is long over, the geopolitical mindset governing the distribution of aid has not changed very much. Old recipient countries may have been replaced by new ones, but the underlying rationale of using aid to promote donor countries' strategic interests is still very much alive. Instead of allocating their aid based on where it is most needed, rich countries often favor recipients that are of direct political or economic interest to them. As a result, the most impoverished people of the planet actually receive less aid than people living in middle-income countries.
- Many rich nations tie their development assistance to purchases of goods and services from the donor country. Poor countries get aid, but only under the condition that they spend it in a way that benefits businesses in the donor nation.
- Overpriced technical assistance is a form of inefficient aid that is closely linked to tying. In 2003, an estimated $18 billion—more than a quarter of total aid—was spent on technical assistance, mainly on consultants advising and supporting recipient governments. While there is a very concrete need for expertise in poor countries, much of technical assistance is heavily overpriced. In 2002, aid donors spent an estimated $50 to $70 million on 700 international consultants in Cambodia—an equivalent of the salary of 160,000 Cambodian civil servants. (Hirvonen, 2005)*

The system of *tied* aid means that aid money is only provided under certain conditions. For instance, the recipient country or project is forced to use technical advisers and equipment exclusively from the donor country, or to buy supplies from sources authorized by the donor. While this increases the prosperity of the donor country, it does little to help many struggling nations who cannot afford

---

* P. Hirvonen, *Stingy Samaritans: Why recent increases in developmental aid fail to help the poor*, 2005. Found at: http://www.globalpolicy.org/component/content/ article/240/45056.html

the conditions tied to the aid. In a country battling widespread HIV/AIDS, for example, funding for treatment may be tied to the condition that they buy drugs exclusively from pharmaceutical giants at almost five times the price of alternative sources. This would greatly diminish the real value of the aid.

There are further questions beyond who benefits from aid that is tied to any conditions. Is it ethical to provide aid only to countries that will promote the foreign policies of the donor? Should donors be able to dictate how aid money is spent by the recipients? Is development just another name for globalization, or economic colonialism? Between loan repayments, tied aid and trade restrictions that favour donor countries, some estimates have more revenue flowing out of the developing world and *into* the developed world than the other way around (UN, 2012). On a purely practical level, many large-scale aid models simply have not improved the lives of the people they are intended to help in any significant way. Even non-governmental organizations, which should be small and flexible enough to help on a grassroots level, are often hampered by the interests of their sponsors or the agendas of powerful groups who control the political and logistical factors where they are working. The competing interests of stakeholders are always the primary obstacle in a top-down aid system. These are both structural problems and ideological ones.

Analyzing these structural problems for their root causes exposes underlying ideological issues. Wealthy donor countries openly using aid as leverage for their economic and political policies is, in itself, an exercise of power and privilege. It can be argued that the very words *developed* and *developing* as descriptors for nations reveal the paternalism in this ranking system, exposing our institutionalized notions of dominant culture hiding within the feel-good concept of aid. In other words, development funding is not just tied to certain business conditions, but it is also tied to the notion that Western culture is superior because it controls the wealth. Social progress, in this ideology, means emulating the socio-economic models of the developed world. This is the *fine print* beneath the offer of help. In this big-aid model, when economically vulnerable nations let donor countries own their problems, developing countries often lose control over the solutions and perhaps even their identities in the process. But in a globalized, connected world, what is the alternative?

Model 2: Social Action From the Bottom-Up—Barefoot College

In Afghanistan, a village is illuminated at night for the first time in its collective memory because an illiterate grandmother built and installed a network of solar-powered lanterns. She learned how to become a solar engineer at the Barefoot College in Tilonia, India. Women from around the world have received free solar technology training to take back to their villages. Grandmothers, it turns out, make excellent technicians. This is bottom-up social action at work.

The Barefoot College (BFC), founded in 1972 by Sanjit "Bunker" Roy, is a learning environment that fundamentally differs from the classic notion of education. At BFC, the teachers are local people with traditional skills. No university graduates are needed or wanted. Building solar panels is one of many skills being taught at Barefoot College. The curriculum is dictated by poor, rural people, some making less than $1 per day. At Barefoot College, the students and teachers are interchangeable. There are no theory classes, just practical training that uses the know-how and resourcefulness of local people to make their own lives better. The learning technology is kept to a minimum with nothing more than mobile phones and PCs. It is estimated that by 2015, mobile phone connections in India will approach one billion (Watkins et al., 2012), so the cost is kept low and the training is accessible. Solar-lit night school provides literacy training to rural children on *their* schedule, a workable solution in communities where children must contribute to the subsistence of their families during the daytime. The college claims an impressive success rate—over 15,000 women have been trained who now provide 500,000 people with basic services such as healthcare, clean drinking water, and education. The solar-electrification program alone is estimated to save two million litres of kerosene in India every year. There are now 24 colleges built on the Barefoot model, teaching skills without issuing degrees or charging tuition.

Poor, rural men also give and receive training at the college, but the emphasis on equity and equal opportunity for women has been one of the most successful innovations at the college. There is no classism, sexism, or ageism at Barefoot College. The school's philosophy is based on a fundamental respect for the capabilities of local people whose traditional knowledge and resourcefulness are enhanced with aid, not replaced by it. Bunker Roy, whose background is one of privilege, chose to devote himself to this grassroots model as a result of his own social analysis of effective ways to solve social problems in poor communities. Roy states:

> You have a graveyard of successful failures everywhere in the world with this top-down solution that has not worked. With foreign expertise...they don't know the culture and they don't know what's happening in the countries.... There is a growing anger. We have to look for alternatives. It has to be bottom-up, it has to be indigenous, it has to develop solutions from the ground up, and it has to be both community based and community managed. (as quoted in Williams, 2011)

Compare the ideology of this bottom-up approach with the top-down approach of international development and the differences are obvious. The Barefoot model is built upon a principle of respect for the identity and culture of the participants who dictate the curriculum. Projects are small-scale and realistic, and the training is reproducible—it will work in the village as well as the classroom. This type of training is scalable. One trained individual can train as many more as are needed and BFC works with local governments to establish satellite campuses where their graduates can spread their expertise. Technologies such as solar-powered lighting and cooktops, built and maintained by their graduates, can improve lives in entire regions. It is an efficient use of resources with a proven return on the investment in training and basic equipment. Furthermore, control of all resources is kept within the college; any donations from individuals, foundations, or government, are taken on the condition that they are not tied in any way. The success of Barefoot College has made Bunker Roy an international spokesperson for grassroots solutions to social problems and a vocal advocate for the self-determination of indigenous peoples.

Grassroots Activism—Obstacles and Challenges

Bottom-up solutions like Barefoot College face many obstacles and challenges to constructive social action. The most significant challenge is how to work for change within the structures that produced the original problems. Bunker Roy's vision of a college without degrees or diplomas required a strong commitment to that model and a willingness to turn down donations and funding that might subvert the Barefoot plan. Another example of bottom-up action is the Occupy Movement. It first came to widespread attention in 2011 with the Occupy Wall Street demonstration and, like Barefoot College, it is committed to their social action framework, but creating change has proven to be more problematic for them. Occupy is a global network of activist groups protesting social and economic inequality whose slogan, "We are the 99%," reflects their commitment to resist the policies of "the 1%" who control the world's resources. In so doing, they refuse to work within the current political systems of its member countries and regions, which they consider primary contributors to the problem of inequity. While this keeps the structures they are fighting against in clear focus, without the compromises frequently required while working from within a system, it also keeps Occupy out of the mainstream, without access to political resources and allies that could help their cause. "The movement believed that it would be possible to achieve social transformation without really engaging with groups or individuals who had power to help or hinder its cause" (Roberts, 2012). This is a strategy that is grassroots in its rejection of top-down solutions, yet has not maintained the momentum of successful movements before them, like the Women's Rights Movement, which also worked for structural change from within. As the Barefoot and Occupy Movements demonstrate, there is no universal approach for grassroots activism. Some will succeed and others fail—sometimes on the strength of the commitment of the individuals involved, sometimes from factors beyond their control. The constant pressure to compromise, conform, or give up makes grassroots organizing for social change an objective with considerable challenges. And yet, bottom-up social action remains the most accessible vehicle for change for most people.

Model 3: Online Social Action—Activism or Slacktivism?

A friend posts a petition to release a political prisoner in Burma on their Facebook page, and you sign it. Do you know who this prisoner is? Do you even know what's going on in Burma? Or, do you sign it simply to support your friend? Online social action, sometimes called "clicktivism," uses the tools and strategies of online social networking to raise support or apply pressure—what used to be called activism. It is a world where ideology rubs shoulders with marketing, technology, and popular culture. You have to be connected to do it and your support will be recognized in "click-rates," not the offer of a homemade sandwich on the picket-line. Critics of this new style of social action believe it is shallow, insincere, and occupies the lowest level of political engagement. Micah White, calling it the "Wal-Mart of activism," asserts that clicktivists "unfairly compete with legitimate local organizations who represent an authentic voice of their communities." In other words, the people doing the real work, in the real world (White, 2010). Also known as "slacktivism," it has been characterized as activism for a lazy generation who needn't risk a police baton or even a cold by demonstrating outside or handing out flyers (Morozov, 2009). Malcolm Gladwell (2010) describes this perception succinctly when he writes, "Where activists were once defined by their causes, they are now defined by their tools." Are the days of the deeply committed activist willing to work ridiculously hard and risk their personal security to challenge the status quo really over, he asks? Would the Montgomery Bus Boycott,[6] which spawned the American Civil Rights movement, have even take place if it had started as an online petition instead of the arrest of Rosa Parks?

Boycotts, sit-ins, and non-violent confrontations are not obsolete, as the Occupy Movements illustrated, but they certainly have competition from the online protest sites. This worries critics who are skeptical of the easier option to civic participation than on-the-ground activism or even volunteering. Gladwell (2010) states, "The instruments of social media are well suited to making the existing social order more efficient. They are not a natural enemy of the status quo." Online protest, the argument goes, will not change the world.

Defenders of online activism, tell a different story. Ty Richmond, a resident of Jacksonville, Florida, didn't want his daughter attending a high school named after a founding member of the Ku Klux Klan. After he gathered over 160,000 supporters

---

6. The Montgomery Bus Boycott started in 1955 with the refusal of Rosa Parks to vacate her seat to a white passenger on a city bus in Montgomery, Alabama. That coordinated act of defiance was instrumental in changing the social system in the United States under the leadership of Dr. Martin Luther King and has become a symbol of non-violent civil disobedience.

on a change.org petition, the Nathan B. Forrest high school students and school board decided to change its name (Richmond, 2013). It is hard to argue this isn't genuine, grassroots activism when a citizen makes an impact of this kind in his own community. On a larger scale, Avaaz.org, the activism site started by Canadian Ricken Patel, has over 30 million registered members, staff in over 30 countries, and claims to have the attention of world leaders. With an operations budget of $12 million per year from member donations, they can generate an effective social action campaign using marketing techniques and computer algorithms to engage its members (Cadwalladar, 2013). Without the overhead of a physical organization, $12 million will buy a lot of online presence. But are the members of Avaaz really committed? Would they stand outside in the rain?

New media analysts argue that the question is irrelevant. They see the potential for online collaboration as a powerful force for mobilizing human effort. Clay Shirky, one such analyst, considers the passive consumer of the last generation, who absorbs one-way media produced by professionals and beamed through our televisions, to be less socially engaged than today's uploader of holiday pictures and kitten videos. This person, by actively engaging with others online, is using their surplus time in a way that is collaborative, which automatically makes it a potential social resource. Online sharing, according to Shirky, occurs on a spectrum of communal value—the more people who find online sharing useful, the more agency or power it has to change things. Our desire to be social animals, combined with our technology to connect to one another, is erasing cultural barriers and making us producers, not just consumers. That, Shirky says, will empower us to change society. When we share our kitten videos, we're opening the door to sharing our deepest anxieties and discontents about society. When we take the next step of sharing our personal talents and resources to produce change, we collectively have a great deal of clout. The shift from sharing content to sharing ideals and values will naturally lead to collective social action as the tools for collaboration become more accessible to a large part of the world via the Internet. This technology is not creating a new kind of human being, but simply utilizing our tendencies in new ways:

> It is just new opportunities linked to old motives via the right incentives. When you get that right, you can change the way people interact with one another in fairly fundamental ways, and you can shape people's behavior around things as simple as sharing music and as complex as civic engagement. (Shirky, 2010)

Sharing on a huge scale online, says Shirky, is the new frontier of social action. According to this view, a revitalized, "participatory culture," where people have the means, motive and opportunity to get involved in social action, is the exciting potential of the online world.

Model 4: Social Entrepreneurship—"Good" Profit?

Model 4: Social Entrepreneurship—"Good" Profit?

Online activism is creating new avenues for social action, but many of them are simply established methods delivered differently such as signing a petition online instead of signing your name on a piece of paper. Business for profit *and* social good, often called social entrepreneurship, has been hailed as a dynamic model where everybody wins, but critics say it is just straightforward capitalism and that the social dimension is a new way of marketing goods and branding companies.

Social entrepreneurship (also called social enterprise) is typically a smaller-scale version of corporate social responsibility,[7] where large companies conduct business with self-awareness of their environmental and societal impact. But as the Occupy Movements illustrated, citizens are skeptical that corporations can truly act in the public interest. Small- and medium-sized social entrepreneurship is typically viewed as the best opportunity for capitalism with a conscience—businesses that give back to the community. This social action model uses the existing economic and political structures of market capitalism and globalization to do grassroots social action at the same time as making a profit. The **fair trade** movement is considered to be an example of social entrepreneurship as its aim is to conduct international commerce sustainably. However, other self-identified social enterprises have been criticized for appropriating the objectives of social action purely for profit and making very little positive impact. TOMS shoes, for example, which gives away a pair of shoes to a child in need for every pair purchased, has been criticized for building its profit model around impoverished children, using their plight as a sales tactic, and building the cost of the donated pair into the cost of the purchased pair, sacrificing no profit margin. They aren't addressing the root problem of why those children need shoes, and may even be feeding into the structural cause as part of the system that creates such inequities. Supporters of TOMS argue that this is moralizing and that children who need shoes, are getting shoes.

Fashioning Change is another example of a social enterprise, but this online clothing store makes social action the primary role of their business. They call their approach "changeonomics," which they defined as "the concept of leveraging the fiscal power of everyday purchases to drive bottom-up, systemic social change" (2012) on issues such as environmental sustainability, fair wages, and human rights. Fashioning Change asserts that all products on the website have been stringently vetted to meet high standards for human health, protection of the earth, and human rights. They claim that just over 50 percent of suppliers that apply to sell

fair trade

An ethical business model where producers and labourers are paid a living wage and work in safe and humane conditions, and products are made using environmentally sustainable methods.

---

7. Harvard Business School defines it this way: Corporate social responsibility encompasses not only what companies do with their profits, but also how they make them. It goes beyond philanthropy and compliance and addresses how companies manage their economic, social, and environmental impacts, as well as their relationships in all key spheres of influence: the workplace, the marketplace, the supply chain, the community, and the public policy realm (Harvard, 2008).

their goods on the website are rejected because they don't meet their standards for quality and sustainability.

Clearly, this is a social action model that is still evolving. Can business be a tool for creating structural change in society? Can profits coexist with sustainability? Like online activism, this debate on the potential of working with existing top-down structures to create grassroots change is ongoing.

Four Social Action Models—A Summary

Social action for social justice aims to make constructive change for the betterment of humanity. It is a challenging pursuit. Individual acts such as those of Martha Payne, Ricken Patel, and Bunker Roy can have a profound impact on social problems, both locally and globally. The top-down development model distributes resources on a huge scale, but not always equitably or effectively, and powerful interest groups determine its application. Independent, grassroots solutions offer some of the most effective models of social action as they tend to be realistic and doable, and give people agency in their own future. However, grassroots organizations face many challenges to success, sometimes from within their own ranks. Social action must also take place within existing systems. The rejection of the existing system and the refusal to work from within it for change is a defining principle of the Occupy Movement. While powerful in its message, this unwillingness to compromise can create the kinds of practical problems that have plagued the Occupy Movement and contributed to its loss of momentum. The effort to change a system while being forced to work from within it is one of the primary challenges of any social movement. Social entrepreneurship and online activism answer this challenge by embracing the system and using the tools of market capitalism and technology to drive social action. Social entrepreneurship typically marries profits with philanthropy, but in the most progressive models, prioritizes systemic change over profit. All of these models have an online component, ranging from informational websites and social media pages to online stores. As more traditional protest-style activism goes to the Internet and social movements rely on electronic communications to spread their message, online activism may redefine how we collectively approach social problems.

SOCIAL MOVEMENTS

Social action takes many different forms. When social action occurs at the group level, we usually call it a **social movement.** Social movements vary and are sometimes hard to define, but they share certain characteristics. For one thing, social movements tend to bring together people who are interested in advocating for some kind of change. Thus, social movements often have a distinct collective iden-

social movement
A group of people with a common ideology who try to achieve common goals. Social movements can also be described as organized groups of people who may encourage or discourage social change.

tity and are structured around a particular goal. They also have clearly identified opponents. Their goals can either be specific, narrow, or broadly aimed at bringing about social change. Social movements can operate at an individual or societal level and can advocate for minor or radical changes in a society. In the 1960s, a cultural anthropologist named David F Aberle outlined four types of social movements: alternative, redemptive, reformative, and revolutionary (Christiansen, 2009; Conley, 2013; Gerber and Macionis, 2010; Types of Social Movements, n.d.).

Alternative Social Movements

Alternative social movements aim to effect changes at the individual level, but do so in limited ways (Types of Social Movements, n.d.). They target a narrow group of people, and are usually issue-driven (Conley, 2013; Gerber and Macionis, 2010). The alternative movement usually focuses on a singular concern that it seeks to address. An example of this type of movement is the Mothers Against Drunk Driving (MADD) organization. This alternative movement focuses on a specific issue: here, that of drinking and driving (Conley, 2013; Gerber and Macionis, 2010).

Redemptive Social Movements

Redemptive social movements also focus on the individual, but they advocate for more radical change. They tend to bring together people with similar concerns in order to discuss a particular problem, or to change and reform their life. An example of this type of movement is Alcoholics Anonymous, or AA, a support group that brings together individuals wanting to achieve sobriety and thus to change their lives (Conley, 2013).

Reformative Social Movements

Reformative social movements seek minor changes, but focus on society as a whole rather than on the individual. Examples of such a social movement would be the Young Women's Christian Association (YWCA) or the Young Men's Christian Association (YMCA) and the Salvation Army. These organizations respond to the social needs of people in society by providing various programs, financial assistance, and helping people find employment. The aim of these organizations is to bring about gradual yet widespread change.

Revolutionary Social Movements

Revolutionary social movements seek radical changes across the entire society. For a movement to be considered revolutionary it should call for change of the dominant economic system (capitalism) or political structure. A good example of a revolutionary social movement is Occupy Wall Street (Gerber and Macionis, 2010).

SOCIAL JUSTICE AND CHARITY

> The goal of social justice education is full and equal participation of all groups in a society that is mutually shaped to meet their needs…. Social justice includes a vision of society that is equitable and all members are physically and psychologically safe and secure. (Bell, 2007, p. 1)

Social justice and charity are both forms of social action, but their goals and impact are different. Social justice usually involves collective public acts that respond to long-term needs to promote social change in institutions and in society. According to Bell, the aim of social justice is to create an egalitarian society that is based on the principles of equality and solidarity, a society that understands and values human rights and recognizes the dignity of every human being (2007). Social justice is directed at the root causes rather than symptoms of social problems. Charitable organizations provide temporary and immediate relief to people in need. Charity may involve acts by individuals or groups aimed at providing for the immediate needs of others through direct services such as serving food and providing shelter. Services such as homeless shelters, food banks, and emergency aid campaigns are all attempts at responding to injustice and are usually not controversial. Charity addresses the symptoms of social problems.

SOCIAL JUSTICE AND SOCIAL ACTION

Social justice work is challenging. It means raising difficult questions about how society is arranged and thinking about the root causes of such problems as poverty, or homelessness, or racism. To engage in social justice is to ask whether people who are wealthy owe something to those who are poor. Such questions are unsettling to those who believe that their good fortune is a product of their hard work and their abilities rather than the result of unjust social arrangements.

The point then is that social action is not a simple matter. Some forms of social action work towards social justice and other forms come in the way of it. Some forms of social action occur at the individual level like the work of Martha Payne. Others occur at the group level like Occupy Wall Street or the Women's Liberation Movement of the 1960s. Some forms of social action offer temporary solutions to problems (charity), while others address the root causes of an issue (social justice). Exposure to the theory and practice of social action leads to the obvious question of where we, as individuals, fit in. With all the emphasis on action in this chapter, perhaps philosophy can offer some insight.

CONCLUSION

> The unexamined life is not worth living.
>
> — Socrates

Socrates, a self-appointed social activist, wandered around ancient Athens asking people to reflect on the value of their lives. He considered it his civic duty to teach his fellow citizens to be more self-aware of their own biases and prejudices. Critical thinking, starting with oneself, was his enduring message. Athenian society of 2400 years ago was no more ready to hear his message than many people are today. He was executed in 399 BCE for his persistent questioning of the status quo, accused of "corrupting the youth of Athens." The reality is that he was killed for challenging the power and privilege of his fellow Athenians, an undertaking that is just as risky in some parts of the world today as it was for Socrates. It was reported that he died without fear or regret. Could it be that social action starts with the individual, addressing our own issues? Could it be that all social problems are individual problems in group form and that being honest with ourselves is the first stage of social change? Here are some of Socrates' recommendations:

- Find a way to build confidence in our own beliefs, and do not be too easily swayed by others
- Do not follow the crowd passively and do not be afraid to break away from the group
- Do not assume *important* people always know what they are talking about or are worthy to lead us
- Do not assume people in authority are right
- Think logically to avoid being overwhelmed by someone's perceived importance
- Regularly scrutinize what we believe
- Take on the duty of self-reflection
- Practise intelligent non-conformity

Socrates was trying to challenge "lazy assumptions" and find the truth, which he considered a pursuit everyone could share in. He purportedly stated, "I am not an Athenian, nor a Greek, but a citizen of the world."

The concept of global citizenship in an intricately connected world requires an understanding of social analysis, media literacy, identity and values, inequality/equity, and social action. This book has highlighted several key concepts of each topic. It is up to us as individuals and groups to decide how to move forward. By learning how to think more critically, we engage in questioning the world around us and looking for alternative solutions to problems. We can better appreciate how we got here and, perhaps, have a clearer vision of where we want to go.

CRITICAL THINKING QUESTIONS

1. Which approaches to social action might be the most effective moving into the future?

2. What are some of the ways in which you can get involved in taking social action?

3. Do you know of any individuals or groups of people who do not have the same rights as the majority in society? What can you do to change this situation?

4. Are global citizens obligated to take social action?

5. Do you view online activism as an effective replacement for hands-on activism?

REFERENCES

Bell, L. A. (2007). Theoretical foundations for social justice education. In M. Adams, L. A. Bell, & P. Griffin (Eds.), *Teaching for diversity and social justice: A sourcebook* (pp. 3–15). New York: Routledge.

Cadwalladar, C. (2013). Inside Avaaz: Can online activism really change the world? Retrieved from http://www.theguardian.com/technology/2013/nov/17/avaaz-online-activism-can-it-change-the-world/print

Christiansen, J. (2009). Four stages of social movements: Social movements and collective behaviour. In *Research starters: Academic topic overviews.* Retrieved on December 27, 2013, from http://www.ebscohost.com

Conley, D. (2013). Collective action, social movements, and social change. In *You may ask yourself: An introduction to thinking like a sociologist* (3rd ed., pp. 699–725). New York: W.W. Norton.

Easterly, W. (2006). *The white man's burden: Why the West's efforts to aid the rest have done so much ill and so little good* (p. 4). Toronto: Penguin Press.

Fashioning Change. (2012). Retrieved from http://fashioningchange.com/about/vision

Gerber, L. M., & Macionis, J. J. (2010). Studying collective behaviour. In *Sociology* (5th Cdn. ed.). Toronto: Pearson Education. Retrieved from http://wps.prenhall.com/ca_ph_macionis_sociology_5/23/6034/1544830.cw/index.html

Gladwell, M. (2010). Why the revolution will not be tweeted. Retrieved from http://www.newyorker.com/reporting/2010/10/04/101004fa_fact_gladwell?currentPage=all

Harvard. (2008). The initiative defining corporate social responsibility. Retrieved from http://www.hks.harvard.edu/m-rcbg/CSRI/init_define.html

Hirvonen, P. (2005). Stingy samaritans: Why recent increases in developmental aid fail to help the poor. Retrieved from http://www.globalpolicy.org/component/content/article/240/45056.html

Morozov, E. (2009). The brave new world of slacktivism [blog post]. Retrieved from http://neteffect.foreignpolicy.com/posts/2009/05/19/the_brave_new_world_of_slacktivism

Nawaz, M. (2011). A global culture to fight racism. Retrieved from http://www.ted.com/talks/maajid_nawaz_a_global_culture_to_fight_extremism.html

Richmond, T. (2013). Duval public schools: No more KKK high school. Retrieved from http://www.change.org/petitions/duval-public-schools-no-more-kkk-high-school

Roberts, A. (2012). Why Occupy failed: The Occupiers' disdain for everyday democracy brings them dangerously close to their neoliberal foes. Retrieved from http://www.prospectmagazine.co.uk/blog/occupy-failed-adbusters-on-year-anniversary/#.U3BEXfldWSo

Rodney, W. (2012). How Europe underdeveloped Africa. USA: Fahamu/Pambazuka Press.

Shirky, C. (2010). *Cognitive surplus, How technology makes consumers into collaborators.* Toronto: Penguin.

Types of social movements. (n.d.). *Boundless.* Retrieved from www.boundless.com

UN. (2012). Trade imbalances worsening effects of global crisis for developed countries. Second Committee told during discussion on macroeconomic policy questions. Sixty-seventh Second Committee 14th and 15th Meetings General Assembly. Retrieved from http://www.un.org/News/Press/docs/2012/gaef3345.doc.htm

Watkins, J., Kitner, K. R., & Mehta, D. (2012). Mobile and smartphone use in urban and rural India. *Continuum: Journal of Media & Cultural Studies, 26*(5), 685–97. Retrieved from http://dx.doi.org/ 10.1080/ 10304312.2012.706458

White, M. (2010). Clicktivism is ruining leftist activism: Reducing activism to online petitions. Retrieved from http://www.theguardian.com/commentisfree/2010/aug/12/clicktivism-ruining-leftist-activism/print

Williams, G. (2007). Disrupting poverty: How Barefoot College is empowering women through peer to peer learning and technology. Retrieved from http://www.wired.co.uk/magazine/archive/2011/04/features/disrupting-poverty/viewall

Achieved status

A social status that is a result of an individual's work, accomplishments, and/or abilities. (Ch. 6, p. 110)

Activist

An individual who devotes time to work either paid or unpaid to bring about social change. (Ch. 8, p. 153)

Ascribed status

A social status assigned to an individual from birth. It is not chosen and cannot easily be changed. (Ch. 6, p. 110)

Backlash

An adverse reaction to a social ideology that becomes influential such as feminism, or to measures intended to address social problems such as systemic discrimination. (Ch. 4, p. 59)

Back stage behaviour

According to the dramaturgical approach, the behaviour that we exhibit only when alone or around more intimate acquaintances. (Ch. 6, p. 113)

Banana republic

A small state that is politically unstable as a result of the domination of its economy by a single export controlled by foreign capital. (Ch. 2, p. 22)

Barriers

These "policies or practices that prevent full and equal participation in society; barriers can be physical, social, attitudinal, organizational, technological, or informational" (Anzovino & Boutilier, 2015, p. 262). (Ch. 7, p. 139)

Bias

The prejudgment of others in the absence of information about them as individuals or an inaccurate and limited way of perceiving the world or a given situation. A negative bias towards members of particular cultural, racial, religious, and linguistic groups, expressed through speech, written materials, and other media, which harms the targets in many ways. (Ch. 3, p. 30)

Capitalism

A global economic system characterized by the private ownership of the means of production/private property. The capitalists' main aim is to produce goods to sell at a profit by keeping the cost of labour and resources low. (Ch. 3, p. 34)

Cash crops

A crop produced for its commercial value rather than for use by the grower. (Ch. 2, p. 22)

Charity

Aid given to those in need. This could be done on an individual basis or an institution or organization engaged in relief services for the poor and underprivileged in society. (Ch. 8, p. 155)

Class

The "relative location of a person or group within a larger society, based on wealth, power, prestige, or other valued resources" (Kendall et al., 2007, p. 655). (Ch. 7, p. 130)

Cognitive dissonance

The resulting tension one experiences when holding on to two conflicting beliefs or struggling with new information in light of old (Gorski, 2014). (Ch. 4, p. 83)

Commodification

The process of reducing a person, idea, service, or relationship, not usually considered goods into an object of economic value that can be bought or sold in the marketplace. (Ch. 3, p. 30)

Comprador elites

The local business class who derive their wealth and status from multinational corporations and links to imperial and/or former colonial empires. (Ch. 4, p. 79)

Cosmopolitan

Belonging to all the world; not limited to just one part of the world. To be free from local, provincial, or national ideas, prejudices, or attachments. (Ch. 1, p. 7)

Critical media literacy

The ability to analyze and evaluate how media messages influence our beliefs and behaviours. In this process, the viewer is not a passive recipient of these messages, but an active participant who is able to critique the content communicated by media. (Ch. 5, p. 94)

Critical thinking

The mental process of actively and skillfully conceptualizing, applying, analyzing, synthesizing, and evaluating information to reach an answer or conclusion. (Ch. 1, p. 3)

Crony capitalism

An economy where business success is no longer based on open competition in the marketplace but on a close relationships among businesspeople, corporations, and government officials. It is characterized by favouritism in the distribution of legal permits, government grants, special tax breaks, lax environmental policy, and other state intervention in favour of business interests. (Ch. 3, p. 43)

Cultural appropriation

The act of taking on or making use of a non-dominant culture without the authority or right to do so. (Ch. 2, p. 17)

Cultural awareness

An ability to interact effectively with people of different cultures and socio-economic backgrounds. It comprises four components: (a) awareness of one's own cultural worldview, (b) attitude towards cultural differences, (c) knowledge of different cultural practices and worldviews, and (d) cross-cultural skills. Developing cultural competence results in an ability to understand, commu-nicate with, and effectively interact with people across cultures. (Ch. 1, p. 3)

Dependency narratives

Stories that distort the structural basis of inequality and positions the donor as superior and benevolent. (Ch. 3, p. 31)

Discourse

The boundaries within which the topic is understood and talked about. Discourse produces knowledge about topics and therefore can regulate thinking and behaviour into categories of "normal" and unacceptable. Discourse can produce what comes to be seen as "truth" however it does not have to be true, only perceived as such—hence, its link to power. To the Apartheid regime and Ronald Reagan, Nelson Mandela was a terrorist. However, to those that opposed the Apartheid system he was seen as a freedom fighter. Reagan and the white supremacist government saw the struggle for human rights and equity as disruptive to the law and order of that State and the economy. Another example is how we discuss the economy. Capitalism has become synonymous with the term economy. The needs and interests of the wealthy that benefit from this particular economic system become universally understood as the best economic system that will serve the interests of society as a whole. (Ch. 3, p. 30)

Discrimination

An act that has the intent or effect of negatively affecting others based on grounds other than merit or acquired skills. (Ch. 6, p. 114)

Dominant groups

Those "characterized by a disproportionately large share of power, wealth, and social status" (Jianghe & Rosenthal, 2009). (Ch. 7, p. 130)

Equality

Sameness where everyone is treated the same without consideration of individual needs, circumstances, background, or history (Fleras, 2012). (Ch. 7, p. 126)

Equity

Promotes the differential treatment of individuals based on need, taking into consideration circumstances, experiences, background, history, and so on. Equity is focused on achieving equality in the outcome. (Ch. 7, p. 127)

Essentialism

A perspective that assumes that aspects of our identities are innate. We are born with them and they remain fundamentally unchanged throughout our lives. (Ch. 6, p. 109)

Ethnocentrism

The tendency to believe one's own culture is superior and evaluate all others in comparison. (Ch. 6, p. 119)

Fair trade

An ethical business model where producers and labourers are paid a living wage and work in safe and humane conditions, and products are made using environmentally sustainable methods. (Ch. 8, p. 164)

Front stage behaviour

According to the dramaturgical approach, the behaviour that we exhibit when in public or around less familiar acquaintances. (Ch. 6, p. 113)

Gender

The roles of masculinity and femininity that we feel or are expected to play—to perform—based on our sex. (Ch. 6, p. 115)

Gender inequality

Unequal perceptions, treatment, and status of groups based on their gender category. (Ch. 6, p. 116)

Global citizenship

A concept based on social justice principles and practices that seeks to build global interconnectedness and shared economic, environmental, and social responsibility. (Ch. 1, p. 2)

Globalization

The increasing integration of world economies, trade, products, ideas, norms, and cultures in ways that affect all humanity as members of the global community. (Ch. 1, p. 4)

Grassroots organizations

Social activism at a local or community level. Grassroots organizations tend to work through existing political structures to promote social change. (Ch. 8, p. 155)

Hegemonic discourse

A way of framing issues that becomes so embedded in a culture that it appears silly to ask "Why?" about their assumptions. Commonsense assumptions predetermine answers, and also influence the questions that can be asked. (Ch. 4, p. 68)

Hegemony

Dominance is not achieved through direct authoritarian rule, but through a process of building consent through social practices where the ruling classes present their interests as the general interests of the society as a whole. Since they rely on consensus building, dominant ideologies and discourses are open to contestation. When resistance becomes a problem for powerful groups or in times of crisis, the moral and intellectual leadership of "experts" is not enough and hegemonic processes are replaced temporarily with "repressive state apparatuses" such as the police and the prison system as was demonstrated by the treatment of protesters during the 2010 G-20 protests in Toronto. (Ch. 3, p. 37)

Ideology

A systematic set of beliefs, perceptions, and assumptions that provide members of a group with an understanding and an explanation of their world. Ideology influences how people interpret social, cultural, political, and economic systems. It guides behaviour and provides a basis for making sense of the world. It offers a framework for organizing and maintaining relations of power and dominance in a society. (Ch. 3, p. 30)

Inclusion

The act of establishing an environment that fosters diversity where all members of that society are believed to be equally valued contributors and participants (Anzovino & Boutilier, 2015). (Ch. 7, p. 124)

International aid (also known as foreign aid or development)

It is the international transfer of resources—money, goods, expertise—from a country or large organization (such as the World Bank), to a recipient country and its people in order to help them emerge from poverty. This stated intention of this large-scale aid is economic development, medical aid for disease control or prevention, or emergency humanitarian relief following natural disasters and the aftermath of war. By some definitions, military aid is also considered international aid when it protects populations from genocide, or helps stabilize a failing state, but this aspect is contentious. (Ch. 8, p. 155)

Interpersonal skills

Personal attributes that enable someone to interact effectively and harmoniously with other people. (Ch. 1, p. 8)

Intersectionality

The experience, or potential experience, of multiple forms of discrimination based on the intersection of different social statuses. (Ch. 6, p. 114)

Intersex

A term used to refer to people whose biological sex characteristics do not fit into the typical definitions of male or female. (Ch. 6, p. 115)

Looking-glass self

The theory that our ideas about our identity are formed through the way we imagine we are seen by others. (Ch. 6, p. 112)

Media consolidation

The process of concentrating the ownership of media outlets by a small number of large corporations. (Ch. 5, p. 96)

Media regulation

Government control of mass media through laws for the purpose of protecting the public interest or promoting competition among media outlets. (Ch. 5, p. 96)

Media representation

The way in which the media portrays particular groups, communities, experiences, ideas, or topics from a particular ideological or value perspective (University of Minnesota, n.d.). (Ch. 5, p. 98)

Media text

Refers not just to words, but also images, sounds, video, taken as a whole message. (Ch. 5, p. 95)

Minoritized groups

This term results from discriminatory discourses about oppressed social groups that become commonsense assumptions that circulate within the larger population and can also be internalized by members of the targeted group. This process limits how individuals are defined and even how they define themselves. A minoritized person is seen not in their totality as an individual, as a representative of a marginalized social group; an erroneous assumption is made that this group is homogeneous. (Ch. 3, p. 36)

Misogynistic

A person or organization that hates, distrusts, or mistreats women. (Ch. 8, p. 155)

Multiculturalism

The "practice of creating harmonious relations between different cultural groups as an ideology and policy to promote cultural diversity" (Anzovino & Boutilier, 2015, p. 3). (Ch. 7, p. 124)

Neocolonial

Structural arrangements where former colonial powers and current dominant powers exercise political and economic control over territories seized during colonialism for corporate exploitation of resources and labour. (Ch. 3, p. 31)

Neoconservativism

An ideology that subscribes to similar economic policies as neoliberalism, but uses divisive politics such as racism, sexism, homophobia, the War on Terror, Islamophobia, and immigration to enlist support for its economic vision and social policies (Smith, 2004). (Ch. 4, p. 58)

Neoliberalism

An ideology which is premised on the right of individuals to compete in the capitalist marketplace to acquire consumer goods and wealth. It advocates that human well-being is best achieved by creating conditions for entrepreneurial freedoms and skills within a system of free markets and free trade supported by the state (Harvey, 2005). (Ch. 4, p. 57)

Norms

Social expectations about attitudes, values, and beliefs. (Ch. 4, p. 55)

Online activism

Also known as cyberactivism or digital activism, uses Internet-based communication techniques to manage and promote activism of any type. While most modern activism has an online component like a website, online activism is conducted primarily through the Internet, where it uses social media, social networks, and the global reach of the Internet to disseminate a message, gather followers, and conduct social action, such as fundraising, organizing petitions, boycotts, and applying pressure to business and government. (Ch. 8, p. 155)

Patriarchy

The systematic privilege and entitlements conferred upon men over women and children through social, economic, and political control. (Ch. 4, p. 56)

Power

The ability to construct representation of ideas or groups through the organization of meaning (e.g., whether one describes a particular armed person as a terrorist or a freedom fighter). It is also the organization of concepts according to cultural conventions within specific contexts that regulates meaning that is used to normalize, conceal, and distort oppressive and regulatory practices. Also, organizing and justifying ideas that groups of people hold about themselves and the world— usually to maintain the ideologies and worldviews of dominant groups. (Ch. 3, p. 30)

Prejudice

Preconceived negative opinions about individuals or groups. (Ch. 6, p. 114)

Privilege

The advantages that are awarded to those with social identities that have benefits which minoritized social identities do not. Those with privileged social identities are usually not aware that they have advantages others do not. For example, heterosexual people can discuss their dates and romantic partners in social situations whereas those that are not heterosexual may be wary of such disclosures for fear of ostracism, being fired, or even physical harm. Those with privileged social identities are seen as individuals whereas those who are minoritized tend to be viewed as representatives of their social identities. (Ch. 3, p. 30)

Race

The "socially constructed classification of human beings based on identified or perceived characteristics such as colour of skin and informed by historical and geographical context; it is not a biological classification. It is often the basis upon which groups are formed, agency is attained, social roles are assigned and status is conferred" (James, 2010, p. 285). (Ch. 6, p. 117)

Racialized

The process of creating, preserving, and communicating a system of dominance based on race through agencies of socialization and cultural transmission such as the mass media, schools, religious doctrines, symbols, and images. (Ch. 4, p. 55)

Reflective practice

The capacity to reflect on action so as to engage in a process of continuous learning. (Ch. 1, p. 12)

Roles

The social and behavioural expectations assigned to different status categories. (Ch. 6, p. 110)

Sex

A term used to describe the biological and anatomical differences between females and males. (Ch. 6, p. 115)

Sexuality

An individual's sexual preferences and orientation. (Ch. 6, p. 116)

Social action

Individual or group behaviour that involves interaction with other individuals or groups, especially organized action towards social reform. Social action is an action by a group of people directed toward a better societal end. (Ch. 8, p. 152)

Social constructionism

A perspective that argues that our identities are the product of the interplay between individual, cultural, and social structures. (Ch. 6, p. 109)

Social democratic

A political perspective that includes aspects of liberalism and socialism. Adherents believe government should promote the collective good and play a productive role in the economy to bring about greater equality and distribution of resources and increase democracy. Government should have public control (not ownership) of the means of production, to meet social needs rather than individual profit, regulate the market to bring about greater equality and distribution of resources, increase democracy in political and economic areas, and protect the environment through central government planning. (Ch. 3, p. 35)

Social entrepreneurship

A commerce model that combines the principles of business with the objectives of social action or charity. (Ch. 1, p. 4)

Social inequality

Difference in the treatment of people on the basis of class, gender, age, ability, race, ethnicity, or citizenship. Generally involves restricting people's full participation in society and limiting resources and opportunities, affecting quality of life (McPherson, 2013, p. 112). (Ch. 7, p. 125)

Social justice

"Full and equal participation of all groups in a society that is mutually shaped to meet their needs. Social justice includes a vision of society in which the distribution of resources is equitable and all members are physically and psychologically safe and secure. We envision a society in which individuals are both self-determining (able to develop their full capacities), and interdependent (capable of interacting democratically with others). Social justice involves social actors who have a sense of their own agency as well as a sense of social responsibility towards and with others and the society as a whole" (Adams et al., 1997, p. 3). (Ch. 7, p. 127)

Social movement

A group of people with a common ideology who try to achieve common goals. Social movements can also be described as organized groups of people who may encourage or discourage social change. (Ch. 8, p. 165)

Social status

The position a person has within a society's hierarchy. (Ch. 6, p. 109)

Social stratification

Refers to "the hierarchal arrangement of large social groups on the basis of their control over basic resources" (Kendall, 2010, p. 214). (Ch. 7, p. 130)

Social structures

The network of social relationships created among people when they interact with each other, within societal institutions according to their statuses in that society. (Ch. 4, p. 58)

Socialization

The process, through interactions with others, by which we come to understand different social statuses and the roles, or behavioural expectations. (Ch. 6, p. 110)

Stereotypes

Beliefs held by individuals about the presumed physical and psychological characteristics of members of a social category. They can be either positive or negative and when applied so generally, individual differences are not recognized, or even defined. (Ch. 3, p. 30)
Unfounded and unwarranted generalizations about particular groups of people. (Ch. 6, p. 113)

Stereotype threat

The effect of negative stereotypes on an individual's performance or behaviour. (Ch. 6, p. 114)

Subsidy

A sum of money granted by the state or a public body to help an industry or business keep the price of a commodity or service low. (Ch. 2, p. 22)

Tokenism

The "practice of including one or a small number of members of a minority group to create the appearance of representation, inclusion, and non-discrimination, without ever giving these members access to power" (Anzovino & Boutilier, 2015, p. 6). (Ch. 7, p. 125)

Trade sanctions

One or more trade barriers that one country places upon another country as a punitive action. (Ch. 2, p. 22)

White supremacy

A belief that white people are superior to all other races. This was reinforced through government policy, legislation and laws, and academic and cultural production to glorify whiteness and malign colour. (Ch. 4, p. 56)

INDEX

Page reference followed by *n* indicates a footnote.